EMBRACING SPIRIT, ENGAGING MINDS

CEM 221/222

Organic Chemistry
Laboratory Manual

2020-2021 Academic Year

Written/edited by Daniel Berger

Table of contents

Required materials for CEM 221/222 laboratory

You are required to have the following materials for CEM 221/222 laboratory:

- This laboratory manual, which contains the experiments we will perform this year.

- Zubrick's *The Organic Chem Lab Survival Manual*, which contains invaluable descriptions of glassware and procedures. You will be required to use this book to prepare for lab, and follow its procedure for keeping your lab notebook.

- A permanently-bound laboratory notebook with permanent page numbering. The bookstore carries the Hayden-McNeil notebook which is strongly suggested for this course. You must turn in copies of your lab notebook pages with your lab reports, and the H-McN notebook makes duplicates as you write.

- A molecular model kit. I strongly suggest the *Molecular Visions* kit sold by the bookstore as it combines cheapness and versatility. You will be expected to identify models constructed with the *Molecular Visions* kit, and to build models from drawn structures.

Expectations for CEM 221/222 laboratory

Preparation and attendance

We will meet for laboratory almost every week of the academic year. You are expected to be present for *every* lab period unless the professor has excused you; however, an excused absence does **not** excuse you from completing the required work. **Limited** time will be available for students who need to make up laboratory work; see the laboratory schedule for the current semester, handed out with the course syllabus and available on Moodle.

Students must turn in a pre-laboratory assignment **before** being admitted to the laboratory, for each week except the first week of each semester and the fall laboratory final exam. Time spent doing the prelab during lab time will **not** be made up, *even if everyone else is allowed to run overtime.* Pre-laboratory assignments are found at the end of each lab procedure.

Failure to complete *all* laboratory work, **including** assignments not associated with experimental work **and** an acceptable report or other written work for each and every experiment, will lower your final grade for the semester.[1] *Reports will not be accepted if they are more than two weeks overdue.*

- If one or two laboratory assignments are incomplete, you will receive a zero, but your grade will not be lowered further.

- If three laboratory assignments are incomplete, you will receive a zero **and** your semester grade will be lowered "half-a-grade," that is, from B- to C+ or C to C-.

- If four laboratory assignments are incomplete, you will receive a zero **and** your semester grade will be lowered by a full letter grade, for example from a B- to a C-.

- If five or more laboratory assignments are incomplete, you will automatically fail the course.

---

[1] Emergencies will require some other arrangement, but you must make those arrangements in cooperation with the professor.

SOME EXPERIMENTS MAY REQUIRE MORE THAN ONE WEEK TO COMPLETE; failure to complete a multiweek experiment may result in the penalty for missing the same number of one-week experiments. For example, not completing a two-week experiment could result in your grade being lowered a full letter grade, rather than the half-a-grade penalty for a single-week experiment.

The goals of any laboratory course are to help you understand material from the lecture and, as much as possible, to introduce you to what goes on in real chemistry laboratories. The goals of this particular laboratory course are to introduce you to basic techniques used in organic chemistry. To get maximum benefit from the laboratory, you need to be free to learn by doing and not hobbled by constant, slavish reference to an unfamiliar text.

Therefore, **you are expected to familiarize yourself with the procedure for each experiment** *before* **coming to the lab**. To encourage this, you have an excellent laboratory textbook (Zubrick), and lab procedures will lack some details. You are expected to interact with the professor and the lab textbook if you have questions on how to perform a procedure.

This is a three-hour laboratory, and the experiments–usually but not always *excluding* instrument time– should take no longer than three hours. While time extensions may be granted at the professor's discretion, you may be cut off at the end of three hours. If your work is not finished, you will receive a zero for the experiment. This is a standard procedure at most universities and colleges.

Some of the experiments will involve the use of instruments to analyze your product. You should take the time to familiarize yourself with the principles behind the instrument and the analysis to be performed. If your text does not provide adequate preparation in this, you will be pointed toward good reference books in the Science Department lobby. The professor will provide training in the use of the instrument.

Often, you will be asked to perform instrumental measurements outside your regular laboratory time. The protocol for doing this will be explained to you.

Again, **laboratory reports or other paperwork are required** as part of each experiment and each laboratory exercise.

You will be penalized 10% of your report score for each violation of safe laboratory procedure[1] AND for each experiment in which you fail to clean up after yourself. Violations that are judged to be sufficiently serious will result in further penalties up to and including being asked to leave and receiving a zero for the experiment.

Laboratory conduct, apart from safety or cleanup violations, will under most circumstances not be taken into account except for adjusting borderline grades; however, egregious violations will be penalized as noted above, or otherwise as the professor deems appropriate.

---

[1] For example, since mercury is a hazardous substance, breaking a mercury thermometer is a lab safety violation, which will be penalized by deduction of 10% of the grade for that experiment. Failure to follow handling and disposal guidelines given in the lab procedures are also lab safety violations.

Preparing for the laboratory

1. In your notebook, write an introduction to the experiment to be performed.

2. In your notebook, compile a table of chemicals to be used, with physical data and a brief mention of their hazard ratings; these can usually be found in their Material Safety Data Sheets (MSDS), available online. Be sure to distinguish between pure substances and mixtures, such as pure NaOH *vs.* 2M NaOH.[1] For example,

Substance	m.p.	b.p.	Hazards
Ethanol, 95%	N/A	78.5°C	Fire: 3 Health: 2 Reactivity: 0
Caffeine	238°C	N/A	Fire: 1 Health: 2 Reactivity: 0

3. Complete the pre-laboratory assignment. Each experiment has a pre-laboratory assignment, given at the end of the experiment description. *The pre-lab is your admission ticket to lab.*

4. Use Zubrick's *The Organic Chem Lab Survival Manual.* It discusses just about everything you need to know about doing organic chemistry lab.

    a. Zubrick has a whole chapter on lab safety. We will follow most of his suggestions and directives.

    b. Zubrick has a whole chapter on how to keep your lab notebook. Follow it!

5. Use the resources in the Shoker Science Center lobby. We don't keep textbooks and laboratory manuals there for show. We intend that you use them to prepare for the laboratory, so that you have a better idea of what to expect. See also the section on reference materials, below.

6. Procedural steps are to be written **as or after they are performed** (see the next section). Do not write them in advance! You have the procedure in the lab manual to refer to.

Laboratory notebooks

Laboratory notebooks must be *permanently bound* and have *permanently and consecutively numbered pages.* An appropriate notebook is one of the required course texts. You will hand in **copies** of your lab notebook pages and of any instrument readouts (not the originals!) with each lab report. Notebooks will also be inspected at random intervals, during lab time, to ensure that you are keeping your notebook properly and writing in it as you work.

In addition to the notebook-pages portion of each report score, your notebooks will be graded separately twice during the fall semester: during fall break, and during finals week. These notebook grades will each be worth 25 laboratory points. Notebooks may or may not be graded during the spring semester at the professor's discretion.

---

[1] If you don't know what "2M NaOH" means, you need to review General Chemistry!

For guidance in writing your laboratory notebook, **you are expected to follow** the chapter on notebooks in Zubrick, though we will deviate from Zubrick in some things. The touchstone for a good laboratory notebook is this: *someone who has never performed the experiment should be able to reproduce your procedure and understand both your results and the purposes of the experiment.* Here are a few guidelines, based on standard practice:

- It is strongly recommended that you diagram your experimental equipment before lab, in your notebook, as shown in Zubrick Chapter 2. This will help you remember how to set up and run the experiment. Experimental setups are discussed in Zubrick.

- Laboratory notebooks are kept in **ink**, not in pencil.

- Strikeouts are performed by drawing a single line through the material to be deleted, ~~like this~~. This allows an auditor to know that you haven't been cheating.

- **All lab notebook entries must be dated**. If you make several entries on the same day (as you will during lab), you need not date every separate entry. But each page must be dated, and if you make entries on the same page on two different days, the page must show the date of each day's entries. Usually this is done by putting the original date at the top of the page, and then writing the new date above the new day's entries.

- Manipulations and observations are written *in your notebook* **as you work**. In real-world research labs, notebooks are updated every time there is a break in the procedure being performed. Develop the habit of writing everything in your notebook as soon as you do it; for example, take your notebook with you to weigh something. If you are caught writing data (for example, the weight of something) on a separate sheet of paper, it may be taken away from you and destroyed.

  Some students have been in the habit of writing data in their lab manuals rather than in their notebooks. *Students caught doing this will be penalized 5 points per incident.*

  If I see that you are not keeping up with what you are doing in your lab notebook, your notebook score for that experiment will automatically be zero points of five. I have worked in a lab before; I know how it works. You will not be penalized for not writing while you are manipulating things; nobody has more than two hands. But you *will* be penalized if I see you "doing nothing" and your notebook is not up-to-date.

- All spectra and other printouts of instrument data must be fastened into your notebook. It should be made clear to the reader where to find the printouts, if they are not on the pages immediately following the experiment. Occasional "surprise" notebook inspections will be carried out to ensure that you are not handing in your only instrument readouts with your reports.

- A conclusion should be written as soon as all data are in.

Reports

Reports will be *entirely* typewritten or word-processed[1] and patterned on the format for a paper submitted for publication in the chemical literature; see the *ACS Style Guide*, on reserve in Musselman Library.[2] Except for the abstract, the report will be **double**-spaced and will have numbered pages. You may use ALL CAPS, *italics* or **boldface** type for titles; it is strongly suggested that you take advantage of the "Heading" styles in Microsoft Word. Reports that do not use the appropriate format will be returned to you for rewriting.

You should scan a few papers from the *Journal of Organic Chemistry*, which is available via OhioLink through the Musselman Library website, to get a feel for the format I expect.

Pay attention to the instructions in your lab procedure and given to you orally; some reports will not require all of the sections specified below.

Each report will have

- A cover page, which will include
  - ✓ the title of the report,
  - ✓ the author's name,
  - ✓ the *due* date,
  - ✓ the class and professor,
  - ✓ The Pledge, signed:
    I certify this lab report complies with the honor system guidelines for lab reports. Specifically I certify that (1) the data in this report were collected by myself; (2) I have learned how to perform and can reproduce the analysis by myself; (3) I have not utilized reports nor other aid from previous sections of this class; (4) my writing does not plagiarize the work of others.
    and
  - ✓ a single-spaced, left-justified **Abstract**[3] of the contents of the report.

- An **Introduction**, in which the problem is presented and relevant background information is given.

- A body–often, but not always, in two parts: **Results** and **Discussion**–in which the work performed is discussed and the results evaluated. This does *not* consist in simply rehashing

---

[1] Pictures and figures may be **neatly** hand-drawn, but for a neater look see the note on available free drawing software, below.

[2] Please note that the style for the chemical literature is somewhat different than for the physics literature, so that the report format expected for organic chemistry is different than that for physics.

[3] The abstract **must** contain:
- The purpose of the experiment ("Anhydrous thanatolamine was prepared")
- The results of the experiment ("the unknown was identified as bat guano" "a 0.1% yield")
- Anything else relevant.

The purpose of the abstract is to whet the reader's appetite (so s/he will read the rest of your work) and, as a last resort, to give the gist to the reader who simply doesn't think s/he has time to read the whole thing.

the contents of the **Experimental** section, unless such is relevant to a discussion of the principles guiding the work, or of the results. *Do not forget to discuss topics requested in the lab procedure.*

Do not present me with a "calculations" section; normally you just tell me the results. If you have a number of results to be compared, *do* tabulate them rather than just listing them. This shows that you have thought about how the numbers relate to each other.

- A **Conclusion**, in which the results and their interpretation are *summarized.*

- An **Experimental** section, in which details of the experimental or other procedure are given. This is normally placed either after the conclusion or after the introduction. This section seems to give students fits; guidelines and examples are given below.

  The Experimental section is the <u>**only**</u> place in the report where actual masses, as opposed to relative yields, should be given. See the examples given below. **Note:** in the experimental section, both actual masses and relative yields are given, e.g. "the solid product massed 4.23 g (87%)." Elsewhere in the report, only the relative yield is given, e.g. "the product was produced in an 87% yield."

- A list of **References**, placed as *endnotes* and executed in ACS format. See the *ACS Style Guide*, which is on reserve in the library.

- *Copies* of the relevant pages from your lab notebook. **These are NOT part of the report**, but a check on how well you are keeping your notebook. **Do not** refer to anything in the lab notebook pages in your report (for example, **do not** refer to "Figure 1" if "Figure 1" exists only in your notebook pages). **The purpose of the notebook pages is as a check that you are keeping your notebook properly.**

- *Copies* of any spectra or other instrument output obtained during the experiment. These must be labeled as **Figures** and *referred to by number in the body of the report.*[1] You may also include structural drawings or chemical equations as **Figures** or **Schemes**. All figures and schemes should be numbered in the order in which they are cited in the text. Figures and schemes are numbered separately (for example, Figure 1, then Scheme 1–not Figure 1, then Scheme 2).

Some reports will not require all of these items; for example, reports in which no instrument readings were obtained will not have instrument readouts. You may be directed to omit other portions of the standard lab report for particular experiments.

Chemical structures and other drawings

- ISIS/Draw, ACD ChemSketch and KnowItAll are standard chemical structure drawing programs, available to students for free download (to find them, do a Google search). ChemSketch is installed on the Science Department computers. If you have problems finding the programs, ask the professor. These programs will help you generate more professional-looking structure drawings but do have a bit of a learning curve.

---

[1] You *must* discuss spectral features that pertain to the identification of your compound(s)!

- All chemical structural drawings and reaction mechanisms **must** be either neatly hand-drawn or generated using a chemistry drawing program such as the three programs listed above. This is to give you practice; you will have to generate such structures on examinations.

- Because of past abuses, you are **not** permitted to copy pictures of other people's drawings into your report even if credit is given. Doing so will result in your report being returned to you ungraded, and late-report penalties will be assessed.

Stylistic considerations for lab reports

Reports are expected to be **concise**. You are not expected to have a certain number of pages, but your discussion must be **complete**.

If you examine the chemical literature, you will find that all articles are written in the *third person, passive voice*. Rather than "I determined the pH" or even "the experimenter determined the pH," the correct form is "the pH was determined," or better yet "the pH was acidic." Copies of American Chemical Society journals such as the *Journal of Organic Chemistry* are available online via OhioLink's Electronic Journal Center, so that you may look at the format and style used for experimental sections in the chemical literature.

In the chemical literature, the **Experimental** section is normally found either after the **Introduction** or after the **Conclusion**. Very long **Experimental** sections are almost always placed after the **Conclusion**.

"COOKBOOKING" IS NOT PERMITTED. "Cookbooking" means a blow-by-blow account of what you did in the laboratory. An example of cookbooking:

> I measured 5.0 mL of concentrated HCl and weighed out 1.0 g of NaOH. I dissolved the NaOH in 100 mL of water in an Erlenmeyer flask. I added the HCl and mixed. I determined the pH using litmus paper.

The way it should be written:

> 5.0 mL of concentrated HCl were mixed in a 125-mL Erlenmeyer flask with a solution of 1.0 g NaOH in 100 mL of water. The resulting solution was acidic to litmus paper.

Most common procedures can be disposed of by one or two phrases. For example, "the solid product was recrystallized from 10% v/v ethanol/water, with a yield of 5.1 g (80%)." "A solution of 10 g 1-naphthol in 50 mL benzene was added, dropwise, over 30 minutes and the resulting mixture was refluxed for one hour."[1] "Solvent was removed by evaporation and the product (b.p. 80-85°/100 torr) was purified by distillation under reduced pressure." In **Experimental** sections, *shorter is usually better as long as no important information is omitted.*

Normally, the experimental section also includes a resumé of results. Here's a checklist for what should be in your experimental section:

- Briefly describe your reaction apparatus; see the examples below. You don't need to tell me that you used a hot plate to heat the stuff, but you do need to tell me that you heated it. You

---

[1] An example of cookbooking that corresponds to the second example: "I weighed out 10 g of 1-naphthol and dissolved it in 50 mL of benzene in a beaker. I took an eyedropper and added the 1-naphthol and benzene to my reaction, a few drops at a time, until it was gone. This took 30 minutes. I took a reflux condenser and attached it to my reaction flask and heated the reaction mixture on a hot plate until it boiled. I boiled it for an hour." See the difference?

don't need to tell me that you "weighed out" so many grams, but you do need to tell me that you placed so many grams of thus-and-such into a 5-mL conical vial, or a 50-mL round-bottomed flask, or whatever.

- Briefly describe what you did. You don't need to tell me you extracted with a separatory funnel, but you do need to tell me that you extracted the reaction mixture with 20 mL of ether, or that you washed the organic layer with 10 mL of 5% NaOH. Don't forget to tell me how long a reaction was allowed to stir, or stand, or reflux, or whatever.

- The absolute and relative yields are given.[1] Normally, write it like this: "… 357 mg (42%) of a white powder" (or "of a clear liquid" or whatever your product looked like.)

- Results of identifying measurements (melting/boiling points, spectroscopy, and other tests) are summarized. *This will be **required** in this class.* Up to three points of seven will be deducted from your experimental section score, if you do not give the yield and basic physical data in your experimental section.

  o Don't forget to give the expected value, if any; usually melting and boiling points are given thus: "m.p. 228°C (lit.[cite source] 235°C)."

All information not obtained by you during the course of the experiment will be properly cited. This includes, for example, the expected value of a melting point. See the *ACS Style Guide*, on reserve in the library.

Sample experimental sections from the chemical literature

Tetraphenylborate Salt of Protonated 1,8-Diazabicyclo[5.4.0]undec-7-ene (TPB-DBUH⁺).[2] Into a 50-mL round-bottomed flask were placed 1.0 mL (6.69 mmol) of DBU and 10 mL of absolute ethanol. An excess of HCl gas was bubbled through the solution. The ethanol was then removed at reduced pressure. To remove residual hydrochloric acid, the residue was dissolved in distilled water and then lyophilized to give a white solid. The chloride salt was dissolved in 10 mL of distilled water and mixed with an aqueous solution of NaBPh$_4$ (2.29 g, 1.0 equiv, in 10 mL of distilled water). The resultant white precipitate was collected and dried in a drying pistol under vacuum over P$_2$O$_5$ using refluxing ethyl acetate. Mp: 198.5-202.0° C dec. ^{1}H NMR (CD$_3$CN, 250 MHz): δ 7.43 (s, N-H, 1H), 7.26 (m, Ph-H, 8H), 7.00 (t, Ph-H, 8H), 6.85 (t, Ph-H, 4H), 3.45 (m, NCH$_2$, 2H), 3.38 (t, NCH$_2$, 2H), 3.19 (m, NCH$_2$, 2H), 2.51 (m, CH$_2$C=N, 2H), 1.91 (m, CH$_2$, 2H), 1.68 (m, CH$_2$, 6H). ^{13}C {^{1}H} NMR (CD$_3$CN, 62.5 MHz): δ 167.0, 164.8 (q), 136.7, 126.6, 122.8, 55.2, 49.3, 39.1, 33.9, 29.4, 26.9, 24.3, 19.8. Anal. Calcd. for C$_{33}$H$_{37}$N$_2$B: C, 83.89; H, 7.89; N, 5.93. Found: C, 82.30; H, 7.87; N, 6.03.

Potassium Hydrogen 4-Sulfo-3-hydroxybenzoate Monohydrate (15).[3] To a magnetically stirred solution of **14** (20 g, 145 mmol) in concentrated H$_2$SO$_4$ (27 mL) heated to 90° C was added dropwise 30% SO$_3$ in concentrated H$_2$SO$_4$ (3 mL). After 12 h of stirring, a precipitate formed and stirring became difficult. The mixture was heated at 90° C further while mechanically stirred for another 1 h. The reaction was then cooled to room temperature and H$_2$O (100 mL) was added to dissolve the reaction mixture. The

---

[1] See "Calculations for Organic Synthesis," below. "Yield" is reported as a percentage in the body of the report, and as both a number of grams and a percentage in the Experimental section. Please do NOT use the phrases "absolute yield," "relative yield" or "percent yield;" the word "yield" is sufficient. **You have been warned.**

[2] Doering, W.von E.; Zhao, D. *J. Am. Chem. Soc.* **1995**, *117*, 3432-3437.

[3] Venkatesan, H.; Davis, M.C.; Altas, Y.; Snyder, J.P.; Liotta, D.C. *J. Org. Chem.* **2001**, *66*, 3653-3661.

mixture was stirred while 25% KOH (32.5 mL) was added dropwise. The resulting precipitate of the title compound was collected by filtration and recrystallized from H_2O (29.36 g, 79%): mp >250° C; ^{1}H NMR (DMSO-d_6, 400 MHz) δ 10.5 (s, 1H), 7.58 (d, J = 8.2 Hz, 1H), 7.4 (d, J = 6.9 Hz, 1H), 7.3 (s, 1H), 3.85 (s, 1H); ^{13}C NMR (DMSO- d_6, 100 MHz) δ 166.98, 153.38, 134.56, 133.36, 127.77, 119.80, 117.56. Anal. Calcd for $C_7H_5O_6K \cdot H_2O$: C, 30.65; H, 2.55; S, 11.68. Found: C, 31.09; H, 2.51; S, 11.9.

Reaction of 6-bromohexanoic acid (7) with boron trichloride.[1] To a stirred solution of 6-bromohexanoic acid (1.2043 g, 6.17 mmol) in dry CH_2Cl_2 (10 mL) at –78°C was added BCl_3 (1 M solution in CH_2Cl_2, 6.50 mL, 6.50 mmol). The clear solution was warmed to 0°C and stirred for 0.5 h. The reaction mixture was cooled to –78°C and excess methanol (3.00 mL, 74.06 mmol) was added via syringe. The solution was warmed to rt, then diluted with excess ether (25 mL) and washed subsequently with sat. Na_2CO_3 (50 mL), and brine (50 mL) then dried with $MgSO_4$. The excess solvents were removed under reduced pressure to give the methyl ester **4a** as an oil (97%). IR (neat): 2944, 2862, and 1739 cm^{-1}; ^{1}H NMR (300 MHz, $CDCl_3$): δ 1.40-1.59 (m, 2H), 1.61-1.69 (m, 2H), 1.81-1.90 (m, 2H), 2.31 (t, 2H, J=8.75 Hz), 3.39 (t, 2H, J=6.97 Hz), 3.65 (s, 3H); ^{13}C NMR ($CDCl_3$): δ 24.3, 27.9, 32.6, 33.7, 34.0, 51.8, 174.1; m/e 210, 177, 74 (base), 59.

An example of an experimental section for this course

Six tea bags were placed in approximately 150 mL of water, then boiled for 15 minutes. After cooling the solution and squeezing all excess liquid from the tea bags, the solution was extracted three times with 20-25 mL portions of dichloromethane.

The dichloromethane was evaporated on a steam bath, leaving behind 0.1206 g[2] of light green crude caffeine (m.p. 225-226° C). The crude caffeine was purified by sublimation, yielding 0.0105 g (9%) of white feathery crystals, m.p. 234-235° C (lit.[1] 235-237° C).

References

1. Windoltz, M., Budavar, S., Stroumtsos, L.Y., Fertig, M.N., Eds. *The Merck Index, 9th Ed.*; Merck: Rahway, NJ, 1976.

Treatment of error

Be sure to distinguish between human error—i.e. knocking over a flask—and experimental error: the uncertainties inherent in any act of measurement, or the failure of a reaction to run. If your reaction fails you must discuss what might have caused it to fail.[3] However, **vague speculation** is discouraged and **will be penalized**. If you don't know, come talk to the professor! Reports will not be docked for honesty in reporting a laboratory goof, except as applicable for safety violations. *However, if you report something that you did not actually perform in the lab, and the professor remembers otherwise, your report will be appropriately penalized.*

---

[1] Dyke, C.A.; Bryson, T.A. *Tetrahedron Lett.* **2001**, *42*, 3959–3961.

[2] Notice that no relative yield is given here because you don't have any idea how much caffeine is actually contained in your tea sample.

[3] **You are wasting your time** trying to explain "low" yields unless you lost a significant fraction of your product to a spill. Typical student yields for these experiments range from 15% to 100%, and even experienced chemists may not get yields over 40% in some experiments. Any experiment in which you get a measurable quantity of the expected product is a success, **no matter how low the yield is**.

Evaluation of reports

Reports will be marked for grammar, style, and so forth as well as content.[1] Clarity is NOT a style issue, and **writing which is not sufficiently clear will result in the presumption that you don't know what you are talking about**.

Reports should be written for an audience similar to that for a scientific paper. The audience for chemical journals is assumed to be sophisticated in chemistry in general – including common experimental techniques such as recrystallization, distillation, obtaining spectra, and so forth–but unfamiliar with the specifics of the work described. You should write your report accordingly: assume that the reader knows how to carry out laboratory procedures in general, but is unfamiliar with the specific experimental sequence and with the theory behind the experiment.

In other words, you don't need to detail the assembly of a distillation apparatus–although you should draw the apparatus in your notebook!–but you do need to tell me that you distilled something. You don't need to tell me how long you waited for your product to crystallize, but you do need to tell me that you purified it by recrystallization, and from what solvent. On the other hand, the length a *reaction* is allowed to run is usually important; see the procedures quoted above.

Submission of reports

Reports in the above format will be submitted to the tray outside my office, after completion of each experiment. Reports are due at 5:00 PM on the due date, which will be specified in the semester schedule. (Typically the due date will be the date of your next lab period.) Reports will be graded according to the rubric below, and returned to you as soon as possible.

Late reports will be penalized 10% per *calendar* day, up to a maximum of 50% per week.[2] This is a strategic decision; you may decide to delay handing in a poor report for a day so you can polish it up. In such a case a late penalty might be a net gain.

No report will be accepted more than two weeks late – that is, three weeks from the date the experiment was completed. You will receive a zero and your course grade will be reduced, according to the information given on page 9.

Unacceptable reports[3] will be returned to you and will be subject to late penalties. Such penalties will begin on the day you are informed that you must rewrite the report. If a report is returned to you for rewriting, the experiment is counted as incomplete until you return an acceptable report to the instructor. You are not excused from the two-weeks-late maximum if you have had a report returned to you for rewriting. For example, if your report is returned to you for rewriting one week after the due date, you must still turn in an acceptable report before the end of the second week from the <u>original</u> due date.

Reports that are obviously unacceptable for formatting reasons (for example, not double-spaced; improper cover sheet) will be returned to you immediately. There is no guarantee that the content will be found acceptable when you return a properly formatted report, but the late clock will continue to run.

---

[1] Some guidelines are available at www.bluffton.edu/~bergerd/classes/writing.html. See also the rubric, below.

[2] For example, a report that is 8 calendar days late will be penalized 60% (50% for a week + 10% for a day).

[3] "Unacceptable" means that the report does not fulfill the requirements set out above, in the lab procedures and in the grading rubric.

Lab report grading rubric for CEM 221/222

Abstract (5 points)

The *Abstract* should be brief but complete, including a statement of what you were trying to do. All important results should be summarized (for example, relative yield; purity; identification of an unknown.)

Introduction and Discussion (60 points)

Introduction and *Discussion* need not be labeled, nor be separate sections, but should both be present. What goes in the *Introduction* and what goes in the *Discussion* is flexible, and you need not duplicate explanations or background material. **Please do not include a "calculations" section.**

Discuss the expectations for the experiment. What were the goals? This is also the place to present a summary of the general theory or principle(s) behind the experiment, **including the reaction mechanism(s).**

Address each result in enough detail to present the theory (if any) behind the finding and to tell whether the desired results were obtained. Evidence obtained by your own experimental work must be used. Spectra obtained should be analyzed in some way, and reactions involved (if any) should be presented and discussed briefly.

In experiments in which this is relevant, correct calculation of the yield will be worth 10 points. Likewise, each mechanism requested of you will be worth 10 points as part of this section.

Conclusion (10 points)

The *Conclusion* must summarize the important results and address each point raised in the introduction *if relevant to the experimental results*. That is, you should not say "the principles of chromatography allowed us to separate the mixture" but if one of the goals was to isolate clove oil, you should say whether you did or did not!

Experimental (5 points)

The *Experimental* section will *concisely* present the operations actually performed in the laboratory; for guidance, see the laboratory manual and consult issues of ACS journals such as the *Journal of Organic Chemistry* via the OhioLink electronic journal center. Normally this section will include all measurements performed (i.e. starting mass, absolute <u>and</u> relative yield, melting or boiling points, spectral data) during the experiment.

"Cookbooking" (defined in the laboratory manual) will result in a score of **zero** for this section. But we walk a fine line; failure to present (*concisely*) all the important operations will also result in a lower score.

If your experimental section looks like one from the professional literature, it's likely to earn full credit.

Yield of product (5 points)

Yields will be <u>independently</u> calculated by the grader, <u>from the data recorded in your notebook</u>. It will be graded on a 0-5 scale with the largest yield getting 5 points. If the grader cannot reproduce your reported yield <u>using data from your notebook pages</u>, you will receive zero points of five for yield.

These points will be awarded for correct identification of unknowns in those experiments in which unknowns are to be identified. Additional points may be used from the "Discussion" section for this.

Style (10 points)

The report must be well written in ACS standard style with scoring on this scale:

0	2	4	6	8	10
Standard English					Concise, well-written
is trampled on					and well-edited

Points to consider include not only clarity but spelling, grammar, punctuation, proper use of paragraphs, and so forth. Good writing style will not make up for poor or nonexistent content.

For each violation:

✓ Information obtained elsewhere than the laboratory is not properly referenced: subtract 3 points.

✓ Standard ACS format not followed in references: subtract 2 points.

✓ Spectra or reaction schemes are not properly referenced in the text: subtract 2 points.

✓ Standard report format not followed: subtract 5 points. This includes a cover sheet with *all* of the required items.

✓ Inclusion of phrases such as "percent yield" or anything similar: subtract 2 points. See "Calculations for Organic Synthesis" in your manual.

No report will receive a grade lower than zero.

Notebook pages (5 points)

These points pertain to the copies handed in with your report. Your twice-per-term notebook grades are separate from this.

0. Notebook has not been kept during lab. Notebook pages are illegible (including "too faint to be read") or not present; not all experimental measurements and procedures recorded in notebook; something in the notebook is "crossed out" in such a way that it cannot be read; notebook is not kept in ink; notebook pages are not dated.

1. Notebook pages pass inspection for items in (0) but are missing introduction, conclusion, or instrument readouts (TLC plates, GC trace, spectrum or spectra obtained); observations show signs of having been rewritten (e.g. "too neat").

2. Notebook pages pass inspection for items in (0-1) but are not sufficiently clear.

3. Better than (2) but not perfect. All items listed in (0-1) are included.

4. Better than (3) but not perfect. All items listed in (0-1) are included.

5. Notebook pages are easy-to-follow and proper conclusions are drawn. Grader is confident of being able to exactly reproduce what you did in the lab. All items listed in (0-1) are included.

Scaling

Because not every report will have every item listed above, most reports will be scaled to 100 points. The number of total points for the report will be reported to you when your graded report is returned.

Calculations for Organic Synthesis[1]

In any experiment, it is the relationship between *chemical quantities* (moles or millimoles) which is important, rather than that between *physical quantities* such as mass or volume. Because there are no "molemeters," we are forced to use scales and volumetric glassware to measure the quantity of a substance. *This should not be allowed to obscure the fact that it is **chemical quantities** that are fundamental.* Only by knowing the *chemical* amounts of the reactants involved in a synthesis can you recognize the stoichiometric relationships between them or predict the yield of the expected product.

It is helpful to regard a chemical calculation as a "conversion," in which a given quantity (using one set of units) is "converted" into the required quantity (in another set of units). This can be accomplished by multiplying the given quantity by a series of ratios used as unit or dimensional *conversion factors. Unit conversions* are carried out by conversion factors whose quotient is unity. (For example, 454 g = 1 pound, so 454 g ÷ 1 lb = 1 and we use a conversion factor of 454 g/lb or 1 lb/454 g.) *Dimensional conversions* convert between different *dimensions*, such as mass, volume and chemical amount. For example, the density of a substance may be regarded as a conversion factor linking mass and volume, and is used to convert a given mass to units of volume, or vice versa. You should already be familiar with some dimensional conversions.

All calculations should explicitly contain *all* units involved, and **should be checked to make sure that units cancel**, leaving only the desired units in your answer. This does not ensure that your answer is correct, but if the units don't cancel properly your answer is sure to be wrong.

The following examples illustrate some fundamental types of calculations that you will encounter.

Chemical amount and mass

The chemical amount (in moles or millimoles) of a pure substance is converted to its mass by multiplying by its molar mass, a *unit conversion factor*. Remember that the molar mass of a substance is obtained by simply appending the unit g/mol to its molecular weight, which is a dimensionless quantity. For example, the mass of 15.0 mmol of butyl acetate (MW = 116) is

$$15.0\,\text{mmol} \times \frac{1\,\text{mol}}{1000\,\text{mmol}} \times \frac{116\,\text{g}}{1\,\text{mol}} = 1.74\,\text{g}$$

The two ratios ("fractions") are conversion factors: the first to convert mmol to mol, and the second to convert moles (of butyl acetate) to grams. Note that the chemical amount in millimoles must be converted to moles before the conversion factor is applied; otherwise the units will not cancel.

Mass can be converted into chemical amount by inverting the conversion factor(s) before multiplying:

$$1.74\,\text{g} \times \frac{1\,\text{mol}}{116\,\text{g}} \times \frac{1000\,\text{mmol}}{1\,\text{mol}} = 15.0\,\text{mmol}$$

Chemical amount and volume

The chemical amount (in moles or millimoles) of a **pure** liquid is converted to volume by multiplying by its molar mass and the inverse of its density. For example, the volume of 25.0 mmol of acetic acid (MW = 60.1; d = 1.049 g/mL) is

---

[1] This is based on a similar section in Lehman, J.W. *Operational Organic Chemistry, 3rd Ed.*, Prentice-Hall: Upper Saddle River, NJ, 1998; pp. 771-773.

$$25.0\,\text{mmol} \times \frac{1\,\text{mol}}{1000\,\text{mmol}} \times \frac{60.1\,\text{g}}{1\,\text{mol}} \times \frac{1\,\text{mL}}{1.049\,\text{g}} = 14.4\,\text{mL}$$

The volume of a solution needed to provide a specified chemical amount of solute is calculated by multiplying the number of moles required by the inverse of the solution's molar concentration.[1] For example, the volume of 6.0 M HCl (which contains 6.0 mol of HCl per liter of solution) needed to provide 18 mmol of HCl is

$$18\,\text{mmol} \times \frac{1\,\text{mol}}{1000\,\text{mmol}} \times \frac{1\,\text{L}}{6.0\,\text{mol}} \times \frac{1000\,\text{mL}}{1\,\text{L}} = 3.0\,\text{mL}$$

Note that concentrations expressed in mol/L and in mmol/mL have the same numerical value (since 1/1 = 1000/1000). Thus a 6.0 M (6.0 mol/L) solution also has a concentration of 6.0 **mmol/mL**; using these units simplifies the calculation considerably:

$$18\,\text{mmol} \times \frac{1\,\text{mL}}{6.0\,\text{mmol}} = 3.0\,\text{mL}$$

Theoretical yield

The maximum amount of product that could be obtained from a reaction is called the *theoretical yield* of the reaction. Theoretical yields can be calculated using *stoichiometric factors* – ratios derived from the coefficients (expressed in moles) of the products and reactants in a balanced equation for the reaction. For example, the stoichiometric factors relating the chemical amount of product to the chemical amounts of reactants in the following reaction

are 1 mol DBA/1 mol A and 1 mol DBA/2 mol B.

Suppose you were trying to prepare DBA (MW = 234.3) starting with 5.00 g of B (MW = 106.1) and 2.0 mL of A (MW = 58.1; d = 0.792 g/mL).[2] You can calculate the maximum amount of product that could be formed from each reactant by converting the given quantity to moles, then applying the appropriate stoichiometric factor:

$$5.00\,\text{g B} \times \frac{1\,\text{mol B}}{106.1\,\text{g B}} \times \frac{1\,\text{mol DBA}}{2\,\text{mol B}} = 0.0236\,\text{mol DBA}$$

$$2.0\,\text{mL A} \times \frac{0.792\,\text{g A}}{1\,\text{mL A}} \times \frac{1\,\text{mol A}}{58.1\,\text{g A}} \times \frac{1\,\text{mol DBA}}{1\,\text{mol A}} = 0.0273\,\text{mol DBA}$$

Since there is only enough benzaldehyde to produce 0.0236 mol of DBA, it is impossible to obtain more than that from the specified quantity of reactants. Once that much product has been formed, the reaction mixture will have run out of benzaldehyde and the excess acetone will have nothing left with which to

---

[1] Remember that the symbol "M" stands for "moles per liter."

[2] Molecular weights and densities may be found in references such as the *CRC Handbook*, the *Merck Index* or the *Aldrich Catalog*. Molecular weights can, of course, also be calculated from the molecular formula.

react. Therefore, benzaldehyde is the *limiting reagent* on which the yield calculations must be based. The theoretical yield of DBA, in grams, is then

$$0.0236\,\text{mol DBA} \times \frac{234.3\,\text{g DBA}}{1\,\text{mol DBA}} = 5.53\,\text{g DBA}$$

Remember that the limiting reagent is always the one that would produce the least amount of product, *not necessarily the one present in the lowest mass*. In this example, benzaldehyde is the limiting reagent, although there is three times the mass of benzaldehyde present as there is acetone.

Relative yield

It is seldom, if ever, possible to obtain the theoretical yield from an organic synthesis. The reaction may not go to completion; there may be side reactions reduce the yield of product; and there are always material losses when the product is separated from the reaction mixture and purified. For example, in reaction of benzaldehyde with acetone, some benzalacetone may be formed as product, reducing the yield of dibenzalacetone. The *relative yield* of a preparation compares the actual yield to the theoretical yield as defined here:

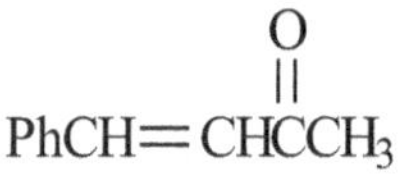

$$PhCH\!=\!CHCCH_3$$
$$benzalacetone$$

that the a by-

$$relative\ yield = \frac{actual\ yield}{theoretical\ yield} \times 100\%$$

If you prepared 4.09 g (the *actual* or *absolute* yield) of DBA using the reaction conditions in the previous section (theoretical yield 5.53 g), the *relative* yield of your synthesis would be

$$relative\ yield = \frac{4.09\,g\ DBA}{5.53\,g\ DBA} \times 100\% = 74.0\%$$

Please, PLEASE **do not** report that "the percent yield was 74.0%"[1] or even that "the relative yield was 74.0%."[2] **Always** report that "the yield was 74.0%" or in the experimental section, "the yield was 4.09 g (74.0%)." Reporting the yield with "gram" units means it's an absolute yield; reporting with a percent sign means that it's a relative yield. Don't overspecify!

Outside the experimental section, only relative yields are to be reported because the relative yield is (in principle) quantity-independent.

---

[1] Say it out loud to yourself to see why this is a problem!

[2] And then there are the students who insist on reporting that "the percent yield was 74.0." WRONG.

Laboratory Procedures
Safety rules

Minimum safety standards and *Disposal* guidelines are given in most experimental procedures and you will, of course, follow them. You are responsible for the safety information in Chapter 1 of Zubrick's *Organic Chem Lab Survival Manual*. You are also expected to follow these general laboratory rules:

1. Proper eye protection, including side shields, will be worn *at all times* in the laboratory. This is a state law. The only exception is when no chemistry or other potentially hazardous work is going on in the area. You are allowed one warning per week; second offenses will draw the 10% penalty for a lab safety violation.[1] **Persistent failure to wear safety glasses will result in permanent expulsion from the laboratory** and consequent failure of the course.

2. Nothing between the shoulders and the knees will be exposed in the laboratory.[2] Open footwear (e.g. sandals or clogs or sneakers with holes in them) will NOT be worn in the laboratory. Students have been sent away from lab to put on appropriate footwear; such lost time will NOT be made up.

   It is recommended that you wear an overgarment such as a lab coat or apron, and that you wear gloves when handling reagents or solvents. Lab coats are available for you to borrow. Dishwashing gloves from the grocery store provide adequate hand protection and do not interfere with dexterity; get a pair that fits snugly. Disposable nitrile gloves are provided (they are expensive; please use no more than one pair per week!) but do not protect as well as a set of dishwashing gloves because they tear more easily. Disposable latex gloves are NOT recommended as organic liquids will pass through them,[3] though they are OK if you are only performing aqueous chemistry.

   After some experiments, your clothes will stink. You should not wear your best clothes to the lab!

3. You will be assigned a bench, a hood and a shared bench. You will be held responsible for cleaning up after yourself. **Failure to do so will result in a 10% report penalty.**

   Students will be assigned responsibility for cleanup of common areas (balance tables, instrument areas and so on) in rotation; failure to clean up these areas will result in the 10% penalty for that student. To avoid upsetting your lab mates, remember that your mother doesn't come to lab with you and clean up after yourself!

4. Any chemical that you do not want to be treated as waste must be CONTAINED (placed in a closed container–this includes stoppered flasks), and LABELED with the **name of the chemical**, the **name of the student**, and the **date**. It should be placed either in a hood or in the drying cabinet, which is vented into the hood exhaust system, unless you are instructed otherwise.

---

[1] If you catch *the professor* in the laboratory without appropriate eyewear during an experiment, you will be awarded 5 points out of 100 on your next lab report.

[2] Short sleeves are allowed; short pants are permitted but not recommended. See also Zubrick, Chapter 1.

[3] A chemistry professor at Dartmouth was killed by wearing only latex gloves while handling a highly-toxic substance; see listings at www.google.com/search?q=dartmouth+chemistry+mercury+death.

Any chemical not contained or properly labeled will be disposed of; this will usually make you unable to complete your work, resulting in a zero for the experiment.[1] The benchtops and drying cabinet will be inspected after each laboratory period to ensure this!

5. You may obtain melting points or perform spectroscopy without the presence of the instructor as long as there is a science professor in the building (*indirect supervision*). You **must ensure** that a professor knows you are in the lab, *especially at night.*

    All other operations except minor cleaning require *direct supervision.* This means that a professor must be in the laboratory with you.

6. No food or drink may be brought into the laboratory.

7. Be careful with glassware and other equipment; it costs money, and replacing it drives up your tuition. Breaking it will drive down your grade, which is partly based on good laboratory technique.

While some of these rules will seem draconian, they are fairly mild compared to the penalties levied by law for improper laboratory procedure. One high school chemistry teacher in California (*ca.* 1998) came within a whisker of doing prison time for washing a quantity of copper sulfate–used to kill algae in ponds and lakes–down the sink. He was let off with a fine and a stern warning because he was found to have acted in good faith. He had not checked state regulations…

Waste disposal

A section on appropriate disposal of wastes from each experiment is included in the instructions for that experiment. In general "mixed wastes" (such as reaction mixtures and solvents from crystallizations) are treated according to standards for the most noxious substance they contain, unless other specific guidance is provided.

- Paper towels used to wipe up chemical spills will be placed **in the hood** until volatile chemicals have evaporated, then disposed of in an appropriate manner according to what is on them. Be sure to use something to **hold the towel down**, or it may be sucked up the flue.

- Chemical waste will be placed in appropriate waste containers. What constitutes an "appropriate" container will be spelled out for each waste substance in the *Disposal* section for each experiment.

  o Never mix halogenated with non-halogenated waste. Disposal costs are much higher for the former, and any amount of halogen makes all the waste "halogenated."

- Glassware will be properly cleaned and placed (normally upside down) on paper towels next to or in your hood. **Substantiated complaints from the next person to use your glassware will result in the 10% cleanup penalty.**

- Odor control is an important part of waste disposal for this laboratory. Many measures you are directed to take are for odor control. For example,

  o Disposal procedures may direct you to "flush" something down the sink. **You must run water continuously for at least five minutes** so that nothing remains in the sink trap.

---

[1] Whether this results in the usual grade penalty for incomplete lab work is up to the instructor.

    o   Sinks, especially sinks that are not in the hood, must be washed carefully after lab to eliminate any chemical odors.

Some definitions related to laboratory safety

Autoignition temperature	The temperature at which the substance will spontaneously burst into flame, given a supply of oxygen. Such a fire can occur if a lid is removed from an overheated vessel, as may happen when oil is heated in a covered frying pan.
Carcinogen	The substance has been found to cause cancer in animals or humans. We will not use known carcinogens.
Caustic	The substance is a strong base. Caustic substances cause burns to human flesh and eat holes in clothing. Not necessarily in common use; see *corrosive*.
Corrosive	The substance is a strong acid. Corrosive substances cause burns to human flesh. They also break down certain fibers (including cellulose, the polymer found in cotton). The term "corrosive" is also used for *caustic* substances in e.g. NFPA hazard ratings (see below).
Dose-response curve	When plotting dose versus physiological effect, the result is normally not a straight line. Instead, at low doses there is often no effect at all (or a therapeutic effect), while at sufficiently high doses the effect is typically toxic. See *LD_{50}*.
Flammable	See *inflammable*.
Flash point	The temperature at which the vapor above a liquid forms an explosive mixture with air. Ignition results in a *flash* (or explosion) rather than a flame. Many organic liquids (such as gasoline) are used at temperatures well above the flash point; this is relatively safe because the liquid's vapor pressure is so high that oxygen is displaced from the air above the liquid.[1] And it still takes a spark to start things off. So don't smoke around gas pumps!
"Flush down the sink"	Pour down the drain, rinsing thoroughly with water. Run water after the material for 5 minutes or more, depending on how much reagent was discarded and how concentrated it was.
Inflammable	The substance can burn ("is able to inflame"), usually meaning that it is likely to burn. See also *flash point* and *autoignition temperature*. Compare the word "inflammable" to an "inflamed" mosquito bite.
Irritant	The substance will irritate skin, eyes or mucous membranes. The fumes may cause you to sneeze.
LD_{50}	The dose that caused death in 50% of a test group (usually of rats). The dose is expressed in mg/kg, milligrams of substance per kilogram of test animal. The *dose-response curve* typically rises sharply at this point, so that amounts smaller than the LD_{50} are often innocuous, and larger amounts are definitely toxic.
Mother	The liquid left over after recrystallization. This is the solution that you dropped your

---

[1] Fuel-air explosives, and internal-combustion engines, are carefully calibrated to mix just the right amount of fuel with air. Too much fuel in the air and the bang is MUCH smaller.

liquor	crystals from.
Mutagen	The substance causes genetic damage (mutations) to cells with which it comes in contact. Mutagens in the bloodstream can adversely affect germ cells, causing genetic defects in offspring. Normally we will not use known mutagens.
Pyrophoric	Pyrophoric materials ignite spontaneously in air below about 45° C. They will also ignite nearby *inflammable* materials.
Residue	Stuff left on glassware after a chemical process is complete and the glassware has been emptied.
Smelly	The substance has an objectionable odor. Smelly substances should be handled only in the hood; this includes cleaning glassware contaminated with them. If they need to be weighed, they must be weighed in **closed** containers.
Teratogen	The substance causes developmental abnormalities in unborn babies; therefore pregnant women should not be exposed to it. We do use some suspected teratogens; if you think you might be pregnant you should inform the instructor immediately.
Toxic	The substance is poisonous. Normally we will not use "highly toxic" substances (LD_{50} < 50 mg/kg). "Mildly toxic" is used to describe substances with LD_{50} > 500 mg/kg.

Material Safety Data Sheets (MSDS)[1]

Material Safety Data Sheets are federally-mandated descriptions of the physical properties and safety data associated with any chemical substance or mixture that is produced commercially. You can find MSDS for toothpaste, for laundry soap, for dishwasher detergent, and for sandbox sand.

MSDS have standard sections that cover different aspects of safety information about the particular substance. The format recommended by OSHA was developed by the American National Standards Institute (ANSI). According to ANSI, the information that must be provided includes:

Section 1. Substance identity and company contact information.

Section 2. Chemical composition and data on components.

Section 3. Hazards identification. This section may or may not include NSTA 704 hazard ratings (see below). More of this information is given below, by section.

Section 4. First aid measures

Section 5. Fire-fighting measures

Section 6. Accidental release measures

Section 7. Handling and storage

Section 8. Exposure controls and personal protection

Section 9. Physical and chemical properties

---

[1] MSDS Writer, L.L.C. www.msdswriter.com/learn_writer.cfm, accessed June 9, 2011.

Section 10. Stability and reactivity

Section 11. Toxicological information

Section 12. Ecological information

Section 13. Disposal considerations

Section 14. Transport information

Section 15. Regulations

Section 16. Other information

OSHA recommends but does not require the use of ANSI format. OSHA does require the following categories to be provided on an MSDS:

Section 1. Manufacturer's Name and Contact Information

Section 2. Hazardous Ingredients/Identity Information

Section 3. Physical/Chemical Characteristics

Section 4. Fire and Explosion Hazard Data

Section 5. Reactivity Data

Section 6. Health Hazard Data

Section 7. Precautions for Safe Handling and Use

Section 8. Control Measures

The OSHA categories are subsumed in the ANSI categories. Most MSDS use the ANSI format, rather than providing only the information required by OSHA.

NFPA Hazard ratings

While at Bluffton we have not (yet) adopted the National Fire Protection Association's NFPA 704 hazard diamond system, it is probably the most useful way of quickly classifying chemical hazard levels. Many Material Safety Data Sheets provide NFPA 704 ratings.

The NFPA 704 hazard diamond has four regions. Clockwise from the left (nine o'clock), they are *Health* (blue), *Flammability* (red), *Instability* (yellow), and *Special* (white). Each hazard class carries a zero (no hazard) to four (extreme risk) rating, as detailed below, except for the white "Special" region, which carries a symbol code.

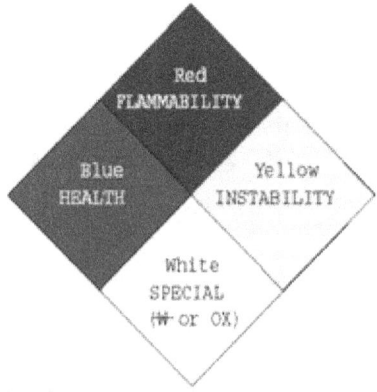

Image source: NFPA website.[1]

---

[1] National Fire Protection Association FAQ, NFPA 704 section. http://www.nfpa.org/faq.asp?categoryID=928, accessed June 9, 2011.

Health ratings (left, blue):

0. Poses no health hazard, no precautions necessary (e.g., water).

1. Exposure would cause irritation with only minor residual injury (e.g., acetone).

2. Intense or continued but not chronic exposure could cause temporary incapacitation or possible residual injury (e.g., ether).

3. Short exposure could cause serious temporary or moderate residual injury (e.g., chlorine gas).

4. Very short exposure could cause death or major residual injury (e.g., hydrogen cyanide, phosphine, carbon monoxide).

Flammability ratings (top, red):

0. Will not burn (e.g., water).

1. Must be heated before ignition can occur (e.g., cooking oil). Flash point over 93°C (200°F).

2. Must be moderately heated or exposed to relatively high ambient temperature before ignition can occur (e.g., diesel fuel). Flash point between 38°C (100°F) and 93°C (200°F).

3. Liquids and solids that can be ignited under almost all ambient temperature conditions (e.g., acetone or gasoline). Liquids having a flash point below 23°C (73°F) and having a boiling point at or above 38°C (100°F), or having a flash point between 23°C (73°F) and 38°C (100°F).

4. Will rapidly or completely vaporize at normal atmospheric pressure and temperature, or is readily dispersed in air and will burn readily (e.g., propane or ether). Flash point below 23°C (73°F).

Instability ratings (right, yellow):

0. Normally stable, even under fire exposure conditions, and is not reactive with water (e.g. sodium chloride).

1. Normally stable, but can become unstable at elevated temperatures and pressures (e.g. propene).

2. Undergoes violent chemical change at elevated temperatures and pressures, reacts violently with water, or may form explosive mixtures with water (e.g. white phosphorus or metallic potassium or sodium).

3. Capable of detonation or explosive decomposition but requires a strong initiating source, must be heated under confinement before initiation, reacts explosively with water, or will detonate if severely shocked (e.g. ammonium nitrate).

4. Readily capable of detonation or explosive decomposition at normal temperatures and pressures (e.g., nitroglycerine, TNT).

Special ratings (bottom, white):

OX   Oxidizer (e.g., potassium perchlorate, ammonium nitrate, hydrogen peroxide). This is an NFPA standard symbol.

W   Reacts with water in an unusual or dangerous manner (e.g., cesium, sodium, sulfuric acid). This is an NFPA standard symbol.

COR   Corrosive, either acid or base. (non-standard; this is covered under health risk level)

ACID   Acid (non-standard; this is covered under health risk level)

ALK   Alkali, which means a base (non-standard; this is covered under health risk level)

POI   Poisonous (non-standard; this is covered under health risk level)

Please note that NFPA 704 hazard diamonds need not be colored – plain white NFPA 704 diamonds are often see on trucks, for example – so you should not rely on the colors to tell you what they mean!

Reference Sources

Physical data of organic compounds may be found in the following sources. All are available in the Shoker Science Center lobby. Details on their use are found in Zubrick, Chapter 3, which also discusses a number of online databases.

- *The CRC Handbook of Chemistry and Physics*

- *The Merck Index*

- *ChemFinder* online database, accessible from the course web page. Caution: mistakes have been found in this database by your instructor!

- *The Aldrich Catalog Handbook of Fine Chemicals*; copies of this are also kept in the lab.

- *Aldrich Library of IR Spectra*

- *Aldrich Library of NMR Spectra*

Chemical safety data may be found in the following sources. All are available in the Shoker Science Center lobby.

- *The Merck Index*

- Hazard.com maintains an extensive MSDS database, from various manufacturers. These may not be fully up to date. Be sure to check the date of the MSDS you are reading!

Discussions of typical laboratory procedures may be found in Zubrick, or any of several organic chemistry laboratory texts in the Shoker Science Center lobby. It is **highly** recommended that you use these references before coming to lab!

Fire extinguishers and inflammable materials

When dealing with inflammable materials, fires do occasionally happen. To prevent and (God forbid!) contain fires, you need to know two things:

- The inflammability of the materials you are working with, as measured by the flash point and autoignition temperature, as well as the chemical reactivity; and

- How to fight a fire involving the materials you are working with.

This section contains basic information about fires and fire extinguishers.[1]

Requirements for a fire

Burning is a surface phenomenon and occurs at the surfaces of objects. Small particles, which have a large surface area per weight, will burn well even if made of "fire-resistant" materials like iron; large objects, like tree trunks, may be hard to light even though they are "inflammable."

A fire requires three things to sustain it: a heat source, a fuel supply, and a source of oxygen. They also require the ability to sustain the combustion reaction, which is a free-radical chain process.

- **Heat Source.** Fire is an exothermic reaction, but it requires an input of energy to start. For example, you need to use friction to start a match; a match to start a Bunsen burner; a spark to start a butane lighter. Normally a fire provides enough heat to sustain itself, but *a fire in the presence of something that absorbs heat efficiently* (e.g. water) will usually die out.

- **Fuel Supply.** Obviously, a fire must have something to burn. This can be something quite unlikely; for example, iron will burn if it is granulated or powdered.

- **Source of Oxygen.** For most fires, this is air. Cut off the air and you cut off the fire. However, Class D inflammable materials (see below) are highly reactive and will extract oxygen from water and even from carbon dioxide. Magnesium flares, for example, will burn underwater or inside blocks of dry ice (frozen carbon dioxide).

- **Fires are free-radical chain reactions.** If something interferes with the chain, the chemical reaction of combustion can be shut down. Some fire extinguishers, notably halon and dry chemical extinguishers, contain "radical scavengers" which disrupt the chain reaction by destroying free radicals.

Types of fires

There are four types of fires, from Class A, which are easy to put out, to Class D, which are NOT.

A. Paper/trash/wood and other more-or-less organic solids. "Ordinary combustibles."

B. Inflammable liquids such as gasoline or paint thinner. Also, hydrophobic organic solids such as naphthalene or stearic acid.

C. Electrical fires, with electricity still flowing to the burning equipment.

D. Burning reactive metals, such as sodium, magnesium, titanium, and so forth. Such metals not only burn at high temperatures but can chemically extract oxygen from water and even carbon

---

[1] For weblinks, see www.bluffton.edu/~bergerd/classes/CEM221/fire.html.

dioxide. Pyrophoric materials[1] such as organoboron, organolithium and organomagnesium (Grignard) compounds are also in Class D because they react violently with water and carbon dioxide. Some reactive metals, such as potassium, are pyrophoric.

In addition, the NFPA defines a special class of fires likely to happen in the kitchen:

K. Kitchen fires. The danger of a kitchen fire is that trying to put it out with water is extremely hazardous. Type K fires are likely to involve burning oil in a very hot pan. Since (a) oil floats on water and (b) the pan is likely to be at a temperature much higher than the boiling point of water if the oil in it has caught fire, using water to put out the fire will splash burning oil everywhere. (The water sinks through the oil, then flashes to steam. The steam expands violently.)

Types of fire extinguishers

Fire extinguishers are normally classified according to the type of fires they are able to handle. You should be familiar, in any situation in which a fire may arise, with (a) the **locations** of nearby fire extinguishers and (b) the **types** of nearby fire extinguishers.

No fire extinguisher can remove the fuel source; an extinguisher can only remove heat or oxygen, or interfere with the chemistry of combustion. Some extinguishers do more than one of these.

Type A Extinguishers

Type A extinguishers are normally colored silver and are suitable *only* for **Type A** fires. They are pressurized-water extinguishers and work by ***removing heat***. The fire's heat goes into heating and evaporating the water, which has a very high heat capacity, and soaking the burning materials with enough water will cool them to below the combustion point. However, all burning materials must be soaked down or the fire will restart. These extinguishers CANNOT be used for Type B fires because burning organic liquids will float on water while continuing to burn. They CANNOT be used for Type C fires because of the risk of electrical shock, NOR for Type D fires because water will support the combustion of Type D materials.

To extinguish a fire with a Type A extinguisher, aim at the base of the fire and soak the burning materials well with water.

CO$_2$ Extinguishers, Type BC

A high-pressure CO$_2$ extinguisher removes oxygen and, to a small extent, heat. The expanding CO$_2$ cools, sometimes enough to produce dry ice snow, but the main effect is to blanket the burning material with a heavy gas that cannot support combustion. CO$_2$ extinguishers leave no residue and so are especially suitable for extinguishing **Type C** fires, which often involve delicate equipment. They are also suitable for **Type B** fires. CO$_2$ extinguishers are NOT suitable for use on Type A fires because the extinguished materials usually retain enough heat to re-ignite when the CO$_2$ dissipates; NOR for Type D fires because CO$_2$ will support the combustion of Type D materials. CO$_2$ extinguishers are normally **red** and have **large nozzles**.

To use a CO$_2$ extinguisher, "spritz" the burning material with CO$_2$ until the fire is out. Watch the materials for several minutes in case they re-ignite.

---

[1] *Pyrophoric* materials will burn spontaneously in the presence of air.

Halon Extinguishers, Type ABC

Halon extinguishers normally contain bromochlorodifluoromethane, a non-toxic, very heavy gas (much heavier than CO_2). This not only displaces oxygen from around the fire but chemically interferes with combustion.[1] Halon extinguishers, like CO_2 extinguishers, are especially suitable for **Type C** fires and delicate equipment, but because they chemically interfere with combustion they are also good for **Type A** and **Type B** fires. However, they are NOT suitable for Type D fires because most Type D combustibles react exothermically with bromochlorodifluoromethane. Halon extinguishers are **red**, and halon is used in the Shoker Science Center's fire-suppression system ("sprinklers").

Halon extinguishers have been phased out because of the damage chlorofluorocarbons do to the ozone layer. However, you may still encounter them; I have one in my kitchen.

To use a halon extinguisher, "spritz" at the base of the burning material until the fire is out. Watch the materials for several minutes in case they re-ignite.

Dry Chemical Extinguishers, Type ABC

The ABC dry-chemical extinguisher, which is suitable for **Type A**, **Type B** and **Type C** fires, is filled with ammonium dihydrogen phosphate ($NH_4H_2PO_4$). This interferes with combustion in a manner similar to halon extinguishers. It also melts at about 350° C, forming a crust that isolates Class A fuel from oxygen. Burning Class D materials can liberate oxygen from the phosphate ion and so this type of extinguisher should NOT be used for Type D fires. Dry-chemical extinguishers are normally red, with small nozzles (and sometimes short hoses).

To use a dry chemical extinguisher, envelop the flames in a cloud of powder by spraying at the base of the fire.

Dry Chemical Extinguishers, Type BC

The BC dry-chemical extinguisher is suitable only for **Type B** and **Type C** fires. These extinguishers are filled with sodium or potassium bicarbonate. The bicarbonate salt interferes with combustion in a manner similar to halon extinguishers. However, enough heat can convert bicarbonate into CO_2 gas and so this type of extinguisher should NOT be used for Type A or Type D fires. This type of extinguisher is better for Type C fires because the residue is much easier to clean up than the ABC dry chemical. Dry-chemical extinguishers are normally red, with small nozzles (and sometimes short hoses).

To use a dry chemical extinguisher, envelop the flames in a cloud of powder by spraying at the base of the fire.

Type D Extinguishers

The cheap version is a bucket of sand, which isolates the **Type D** combustible from oxygen. Sand is silicon dioxide, which is too stable to liberate oxygen even under extreme heat. Commercial Type D extinguishers are called **Metal-X extinguishers** and contain a propellant as well as sand. *These extinguishers are NOT suitable for any other type of fire.*

To use a Type D extinguisher, cover the burning materials completely.

Which type(s) of fire extinguishers are in Shoker Science Center? In your dormitory?

---

[1] Bromochlorodifluoromethane decomposes into chlorine and bromine radicals, which scavenge hydrogen and oxygen radicals essential for keeping combustion going.

CEM 221 Laboratory Final Examination

A laboratory final examination is given in CEM 221 *only*. The lab final is in three, equally-weighted parts:

A. **Fire safety**. This section is based on "Fire extinguishers and inflammable materials." It tests your ability to identify the proper fire extinguisher for a variety of situations, and your knowledge of the basics of fire prevention and fire fighting. This section is multiple-choice and true-false.

B. **Chemical safety**. You have been receiving chemical safety instructions in your lab procedures and discussions all term (including the first lab period, when you signed a safety contract). This section is based on those instructions and tests your ability to identify appropriate procedures for handling a variety of substances and situations. This section is multiple-choice and true-false.

C. **Laboratory equipment**. This section tests your ability to identify and set up the appropriate reagents and apparatus to use for a number of different laboratory tasks. This portion of the exam will be given individually, as a practical, oral examination.

You may be required to pass Sections A and B with a perfect score to complete this course. You may retake the examination as often as necessary for this; but only your first exam score will count toward your course grade.

Molecular structure and your molecular model kit
Structure and Nomenclature

This manual, with pictures and rotatable online models, is available at www.bluffton.edu/~bergerd/Models/. For a short tutorial on using online models, go to www.bluffton.edu/~bergerd/classes/jmol.html.

Thinking in three dimensions is one of the most important skills in organic chemistry. Most organic molecules (especially biological molecules) function through and because of their particular three-dimensional shapes. You must be able to translate flat pictures of molecules into three-dimensional models in your mind in order to do well in either organic or biochemistry. In order to develop this skill, you have been asked to purchase a molecular model kit. This series of exercises, aided by online models, will help you learn to use your model kit.

Using the *Molecular Visions* model kit

The Molecular Visions model kit was chosen for its combination of low price and ability to represent a great many molecules; many professional chemists prefer it as a research model kit. However, like the line-drawing method of representing molecular structures, it is highly stylized.

Three resources will help you learn to use your model kit: the kit manual, which has many color pictures; demonstration by the instructor; and online molecular models, which will allow you to compare the model you build to a more conventional representation. Demonstrations and online models will be provided throughout the course; reading the manual is, of course, up to you!

Some of the laboratory periods will be devoted to learning to use your model kit, and you may be required to identify molecular models during examinations.

Your kit contains several different types of pieces, but the ones that concern us at present are those which allow us to represent organic molecules.[1] These are the *tetrahedral* pieces, the gray *double-bond end-pieces* which go with the gray or red *double bond* pieces, and the *triple bonds*. There are also several *colored balls*.

Tetrahedral pieces should not be confused with the gray double-bond end-pieces! There are black, red and blue tetrahedral pieces, which are the standard (or CPK) colors for carbon, oxygen and nitrogen respectively. To assemble a *tetrahedral center,* take two tetrahedral pieces (usually of the same color) and snap them together. A tetrahedral center represents an atom with four *groups* attached. Remember from VSEPR[2] that a *group* can be either a bond or a lone pair of electrons.

Assemble **tetrahedral centers** from tetrahedral pieces of each color. Notice that the four arms or "bonds" are arranged at angles of 109.5°. The black center represents a carbon atom, with four bonds. The blue center

black   blue   red

---

[1] The gray trigonal pieces will be used later, to represent certain reactive carbon species. Chemistry majors will find the gray trigonal and linear pieces, and the pink pieces, useful when they study *trigonal bipyramidal, square planar* and *octahedral* inorganic and organometallic molecules. The short pink pieces are "bond extenders" used for inorganic and organometallic compounds, which often have longer bonds.

[2] **V**alence **S**hell **E**lectron **P**air **R**epulsion, which says electron groupings (bonds or lone pairs) distribute themselves evenly around a central atom. See Chapter 1 of your textbook.

represents a nitrogen atom, with three bonds and a lone pair of electrons. The red center represents an oxygen atom, with two bonds and two lone pairs.

To represent methane (CH_4), you may use four white balls (representing hydrogen atoms) attached to the black tetrahedral center; ammonia (NH_3) and water (H_2O) can be represented by using three and two white balls, respectively. However, larger molecules will have too many hydrogens for this method to be practical, given your limited supply of white balls. One alternative is to represent a lone pair of electrons by a colored ball, and let blank ends represent hydrogen atoms! Also, divalent oxygen (as in water) can be represented by a *single* red tetrahedral piece.

> *Build models of methane, ammonia and water. Compare them to the online models (www.bluffton.edu/~bergerd/Models/vision2.html). Now build ethane (CH_3CH_3), methanamine (CH_3NH_2) and methanol (CH_3OH); again, compare your models to the online models (www.bluffton.edu/~bergerd/Models/vision3.html).*

Double-bond pieces include several types: *end-pieces,* which are gray; gray *double bonds*; and *half-bonds* which are either gray or red. The gray pieces are used to represent carbon atoms and carbon-carbon double bonds; in order to represent a carbon-carbon double bond, snap a gray end-piece into each end of one of the gray double bonds.

$$\begin{array}{c} H \qquad\quad H \\ \diagdown \qquad\diagup \\ C=C \\ \diagup \qquad\diagdown \\ H \qquad\quad H \end{array}$$

ethene

> *Build a model of ethene,[1] and compare it to the online model (www.bluffton.edu/~bergerd/Models/vision4.html). The "bonds" are arranged at normal trigonal planar angles of 120°. Notice that online models do not always show double bonds! However, you can deduce the presence of a double bond from the trigonal planarity of its two carbon atoms.*

Carbon-oxygen double bonds - *carbonyl groups* - may be represented using the gray and red half-double bonds. To represent a carbonyl group, first snap together one red and one gray half-bond; then snap a gray end-piece into the gray end of the C=O double bond.

Carbon-nitrogen double bonds are rarer in organic chemistry, and the fact that the nitrogen's valence is not filled means that a group may be bonded to nitrogen. To represent a carbon-nitrogen double bond, take a gray double-bond piece. Snap a gray end-piece into one end to represent carbon, and a blue tetrahedral piece into the other end for nitrogen.

methanal   methanal imine

> *Build models of methanal[2] and methanal imine, and compare them to the online models (www.bluffton.edu/~bergerd/Models/vision4.html).*

Triple bonds are represented using the gray triple-bond pieces. **Each piece represents TWO carbon atoms with a triple bond between them and one *open valence* for each atom.** Notice that triple bonds have a normal bond angle of 180°. Carbon-nitrogen triple bonds are also possible (such bonds are referred to as *cyano* or *nitrile* groups), but there is no distinct way to represent them with this model kit.

$$H-C\equiv C-H$$

ethyne

> *Build a model of ethyne[3] and compare it to the online model (www.bluffton.edu/~bergerd/Models/vision4.html).*

---

[1] Commonly known as *ethylene.*

[2] Commonly known as *formaldehyde.*

[3] Commonly known as *acetylene.*

Structure and nomenclature

*Use your models in conjunction with the instructions and the online models. Build the models as you read each description so that you can **see** what is being discussed!*

Alkanes can be represented entirely with the black tetrahedral pieces from your model kit, while alkenes and alkynes require double- and triple-bond pieces to be included. In this module, you will explore the relationship between all-atom representations, line drawings, and molecular models.

Alkanes, alkenes and alkynes

www.bluffton.edu/~bergerd/Models/structure.html

Review the sections in your text concerning **alkane** nomenclature. Using your model kit, build **methane, ethane, propane** and **butane** and compare them to the online models. Rearrange the model of butane to its constitutional isomer, **2-methylpropane** (isobutane). This demonstrates that constitutional isomers use the same atoms, differently connected.

In the table below, line drawings have been used for butane and 2-methylpropane. Line drawings cannot be used for methane and are almost never used for ethane or propane. But line drawings make higher alkane structures much easier to read!

methane	ethane	propane	butane	2-methylpropane

$H_3C-CH_2-CH_3$ (propane)

Alkenes (www.bluffton.edu/~bergerd/Models/struc2.html), of course, must have at least two carbon atoms since the central feature of an alkene is a carbon-carbon double bond. Build **ethene** (ethylene), **propene** (propylene) and **1-butene** and compare them to the models online. Now reconnect your model of 1-butene to represent its *constitutional isomer*[1] **2-methylpropene** (isobutylene).

ethene	propene	1-butene	2-methylpropene

$H_3C-CH=CH_2$ (propene)

Cycloalkanes (www.bluffton.edu/~bergerd/Models/struc3.html) are constitutional isomers of alkenes.[2] Since you need at least three vertices to form a closed figure in geometry, at least three carbon atoms are required for a cycloalkane.

*You should use the odd-colored flexible pieces in your model kit to form small (3- and 4-membered) rings, as **the strain is likely to break the normal, less flexible pieces**. See your kit manual for guidance.*

---

[1] Constitutional isomers have the same formula (for example, C_3H_7NO) but have different "skeletons" (arrangement of atoms along the "main chains").

[2] Both have the general formula C_nH_{2n}.

Build models of **cyclopropane** and **cyclobutane** using the flexible pieces to form the rings. Build **cyclopentane** and **cyclohexane** using the normal tetrahedral pieces.

Notice, from the models you have built with your model kit, how the rings get "floppier" as they get larger. The molecules are shown below as line drawings, in which each vertex represents a carbon atom, and hydrogens are automatically assumed to exist at unfilled carbon valences. You will see that your *Molecular Visions* kit works very much like a line drawing.

cyclopropane	cyclobutane	cyclopentane	cyclohexane

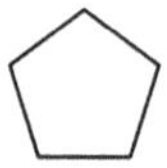

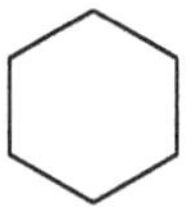

Alkynes (www.bluffton.edu/~bergerd/Models/struc4.html) contain carbon-carbon triple bonds. An important feature of alkynes is their rigid linearity; four atoms are held firmly in a straight line. Build **ethyne** (acetylene), **propyne, 1-butyne** and **2-butyne.** Why is there no other triple-bonded isomer of butyne?

ethyne	propyne	1-butyne	2-butyne

$$H-C\equiv C-H \qquad H_3C-C\equiv C-H \qquad H-C\equiv C-CH_2-CH_3 \qquad H_3C-C\equiv C-CH_3$$

You probably noticed that, while we built 1-butene and 2-methylpropene, we left out one of the isomers of C_4H_8. **2-butene** can be represented in several ways:

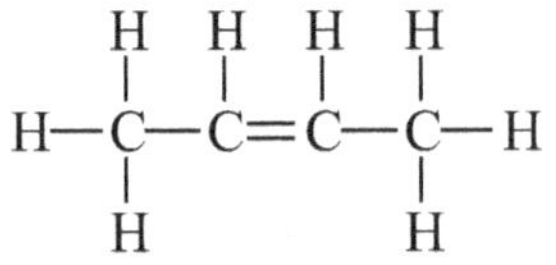

$$H_3C-CH=CH-CH_3$$

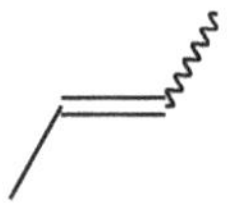

None of these indicates any particular geometry; the squiggly line in the third drawing makes that explicit. However, if you build a model of 2-butene you will quickly notice that there are two possible geometries.

Stereoisomers: Cis-trans diastereomers

(www.bluffton.edu/~bergerd/Models/struc6.html)

The two possible isomers of 2-butene are *cis*-2-butene and *trans*-2-butene; *cis* indicates that substituents are arranged on the same side of the double bond, while *trans* indicates opposite sides. These are *stereoisomers;* the atoms are connected in the same order, but arranged differently in space.

cis-**2-butene**   *trans*-**2-butene**

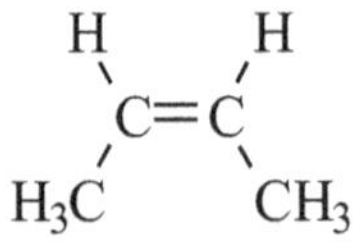

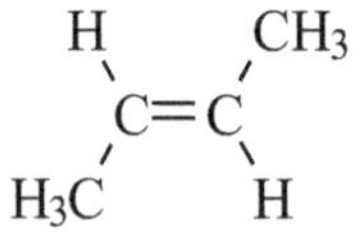

For alkenes, there is a more general way of indicating *cis/trans* orientation: E/Z nomenclature. *Z* stands for the German word *zusammen* (together) and corresponds to *cis*; *E* stands for *entgegen* (opposite) and corresponds to *trans*. The use of these designations is discussed in your text and will be covered in

38

lecture; for our present purpose it is sufficient to point out that another way of naming the two *diastereomers*[1] of 2-butene is

Z-2-butene

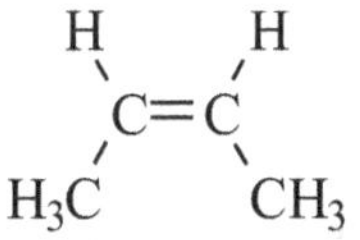

$$H\diagdown C=C\diagup H$$
$$H_3C\diagup \qquad \diagdown CH_3$$

or

E-2-butene

$$H\diagdown C=C\diagup CH_3$$
$$H_3C\diagup \qquad \diagdown H$$

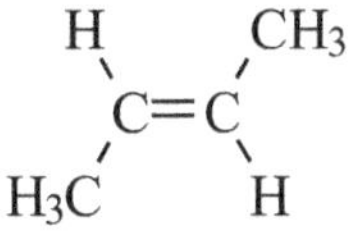

or

In a similar fashion (www.bluffton.edu/~bergerd/Models/struc7.html), cycloalkanes which have two substituents on the ring can have them arranged on the *same* face of the ring *(cis)* or on *opposite* faces *(trans)*. Look at your model of any of the cycloalkanes discussed above, and notice that the hydrogen atoms project both above and below the "plane" of the ring. Now put two methyl groups on the ring at adjacent positions. Is what you have built *cis or trans*?[2]

1,2-dimethylcyclopropane

cis   *trans*

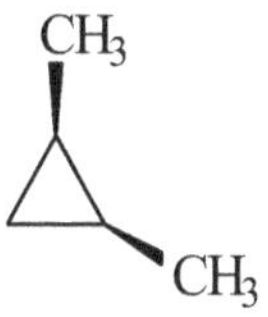

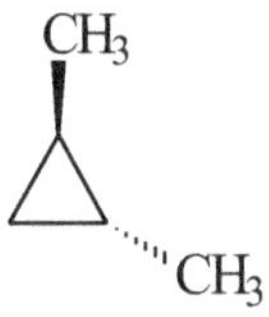

CH_3 CH_3

1,2-dimethylcyclobutane

cis   *trans*

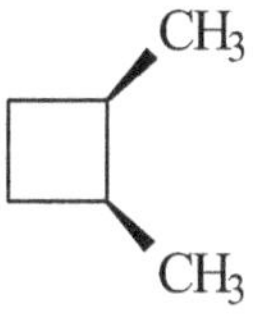

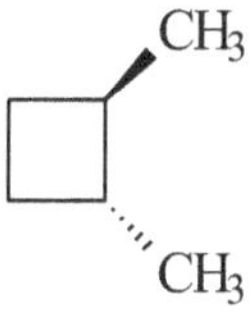

CH_3 CH_3

1,2-dimethylcyclopentane

cis   *trans*

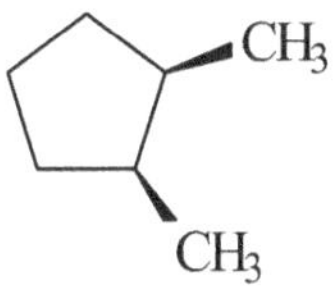

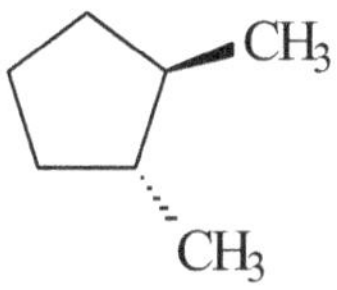

CH_3 CH_3
CH_3 CH_3

1,2-dimethylcyclohexane

cis   *trans*

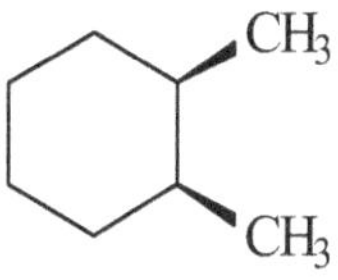

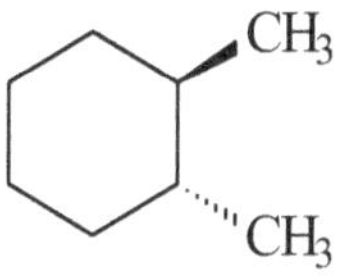

CH_3 CH_3
CH_3 CH_3

---

[1] Diastereomers are stereoisomers which are *not* mirror images of each other.

[2] *Z* and *E* are NEVER appropriate for cycloalkanes!!!

Conformation and Newman projections

(www.bluffton.edu/~bergerd/Models/newman.html)

Use your models in conjunction with the instructions and the online models. Build the models as you read each description so that you can SEE what is being discussed!

Look at your model of ethane. Notice that the central carbon-carbon bond rotates freely. This means that the hydrogens on the adjacent carbon atoms can be either alternating or all lined up; the appropriate terms are *staggered* and *eclipsed*. Since hydrogen atoms have size and thus can interfere with each other, the *staggered conformation* of ethane is lower in energy than the *eclipsed conformation*. The eclipsed conformation has what we call *torsional strain* because of the interactions of eclipsed hydrogen atoms.

The way we emphasize conformation about a particular bond is with a *Newman projection*. To understand a Newman projection, imagine that you are sighting down the carbon-carbon bond in ethane (take your model and do so now!) The carbon atom behind is represented by a large circle; the hydrogens attached to each carbon can be clearly seen in the projection.

staggered ethane

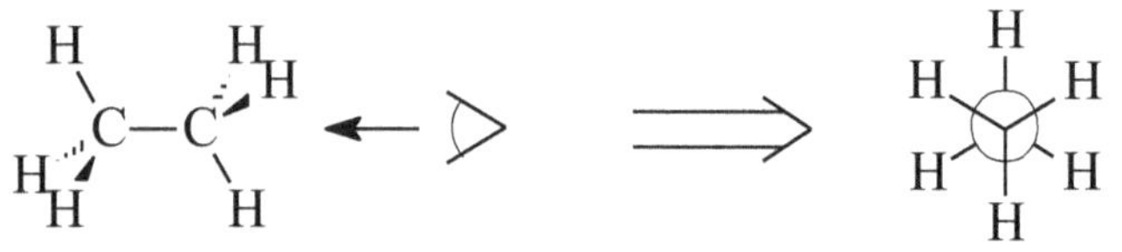

eclipsed ethane

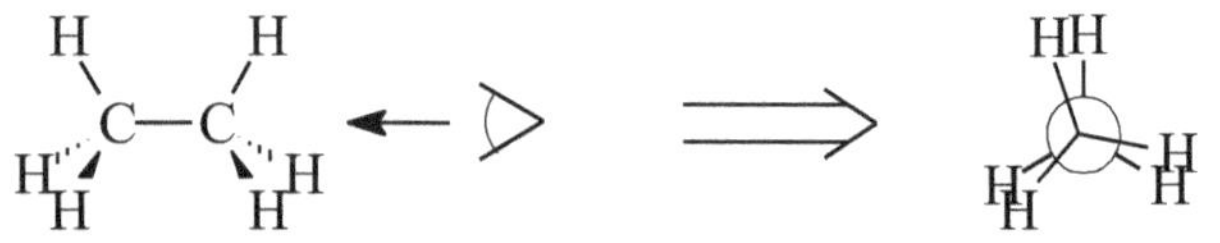

In the same way, propane can be staggered or eclipsed about either of the carbon-carbon bonds.

staggered propane

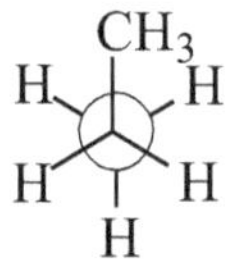

eclipsed propane

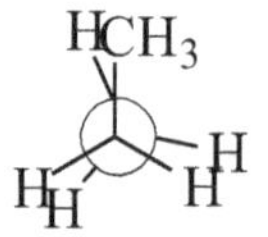

However, butane has another bit of conformational information: when you look down the central carbon-carbon bond, the staggered form can have the methyl groups either adjacent *(gauche)* or opposite *(anti)*. The *anti* conformation is lower in energy than the *gauche*. While this makes little difference in most of the chemistry of butane, conformational considerations become important when we consider cyclo-alkanes.

anti **butane**

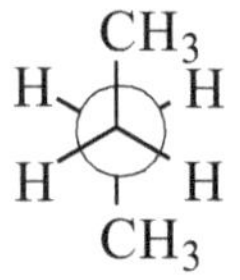

gauche **butane**

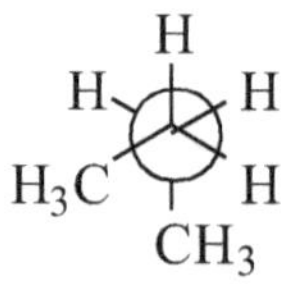

Unsubstituted cycloalkanes (www.bluffton.edu/~bergerd/Models/newman2.html), if planar, would not only have significant *angle strain* (caused by abnormal bond angles in the ring) but also considerable *torsional strain* (from eclipsing of adjacent hydrogens). Cyclopropane and cyclobutane cannot avoid having large amounts of strain in their structures, but cyclopentane and cyclohexane can easily adopt conformations in which not only the bond angles have normal values but eclipsing (and thus torsional strain) is minimized.

Planar *cyclopentane* has almost-normal bond angles of 108°. Nevertheless, to avoid torsional strain the molecule bends into the so-called *envelope* conformation, in which one of the carbon atoms is bent out-of plane. This staggers the hydrogens on that atom relative to those on adjacent carbon atoms.

If *cyclohexane* (www.bluffton.edu/~bergerd/Models/newman3.html) were planar, there would be considerable angle strain as the angles in a planar hexagon are 120° (vs. 109.5°). However, when you use your model kit to build cyclohexane you will see that the ring is really rather "floppy." This floppiness allows both angle and torsional strain to be relieved, and cyclohexane has two main conformational types: *boat* (in which one pair of opposite carbons are bend out-of-plane in the same direction) and *chair* (in which all pairs of opposite carbons are bent out-of-plane in opposite directions).

Boat cyclohexane has no angle strain, but there are several eclipsed interactions between neighboring hydrogens. The actual "boat" conformation is the so-called *twist boat,* in which the eclipsed interactions are alleviated by twisting around some of the carbon-carbon bonds. Nevertheless, the boat conformation is still relatively high in energy because of the unavoidable *flagpole* interactions between the hydrogens on the insides of the "prow" and "stern." The Newman projection below emphasizes the eclipsed interactions along the "gunwales" of the boat.

*The flagpole hydrogens are shown in **boldface**.*

The twist-boat conformation is very difficult to draw; however, it can be seen in the online models (www.bluffton.edu/~bergerd/Models/newman3.html) and you should be able to reproduce it using your model kit.

Chair cyclohexane has neither angle strain nor eclipsed interactions! In fact, it has ***zero strain energy***. (Cyclohexane *does* have gauche interactions.) If you build a model of chair cyclohexane, you will notice that there are two types of hydrogens, depending on whether they point up or down with respect to the ring, or point along the ring "plane." These types are called, respectively, *axial* and *equatorial,* and are color-coded in the on-line version of this tutorial. The Newman projection emphasizes that all hydrogens are staggered; if you examine your model you will see that this is true all the way around the ring.

41

Mark one axial and one equatorial position on your model with colored balls. Now "ring-flip" to the other chair conformation. The colored balls have changed places: the one which was axial is now equatorial, and vice versa.

For unsubstituted cyclohexane, the two chair conformations (or *conformers*) have the same energy. But when we begin to put substituents on the ring, this quickly comes to a halt: because of axial-axial *steric*[1] interactions, most substituents prefer to be in an equatorial position. Very large substituents, such as *t*-butyl groups, *lock* the ring into a particular conformation!

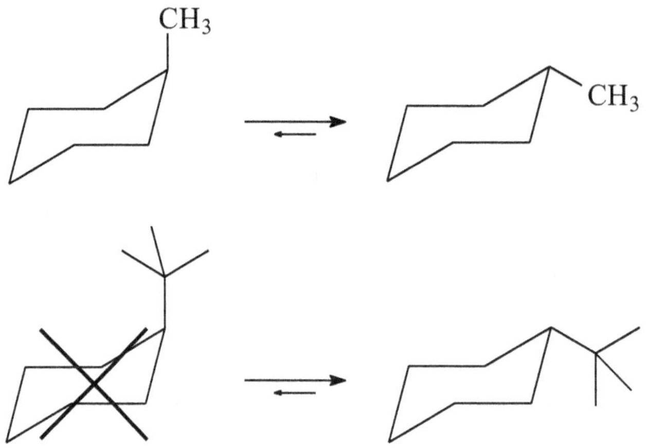

Examine the online models of methylcyclohexane and t-butylcyclohexane in space-filling mode. (www.bluffton.edu/~bergerd/Models/newman4.html) Notice the steric interactions.

When more than one substituent is present, the ring will take on whichever conformation is lowest in energy; if possible, all substituents will be in equatorial positions. However, as you will see, this is not always possible in a chair conformation.

Build models of the following molecules, and report whether one chair conformation will be favored over the other (www.bluffton.edu/~bergerd/Models/newman5.html):

- *Cis-1,2-dimethylcyclohexane*

- *Trans-1,2-dimethylcyclohexane*

- *Cis-1,3-dimethylcyclohexane*

- *Trans-1,3-dimethylcyclohexane*

- *Cis-1,4-dimethylcyclohexane*

- *Trans-1,4-dimethylcyclohexane*

Now build cis-1,4-di-t-butylcyclohexane *(www.bluffton.edu/~bergerd/Models/newman5g.html) . Do you expect it to be a low-energy molecule or a high-energy molecule from steric considerations?*

---

[1] Size-related.

What is observed is that this molecule tends to exist in a **boat** *form because this is the only conformer which allows both* t-*butyl groups to be equatorial!*

Chirality

This manual is available at www.bluffton.edu/~bergerd/Models/chiral.html. Use your model kit in conjunction with the instructions and the online models. Build the models as you read each description so that you can **see** *what is being discussed!*

Chirality[1] is an attribute of objects that makes it impossible to superimpose them on their mirror images. Examples of chiral macroscopic objects include hands, feet, screws, automobiles and so on.

Molecules can also be chiral. Ways of measuring chirality are explained in your text and will be explored in the laboratory; the purpose of this module is to explore chirality using molecular models.

Chiral centers

For a single carbon atom to be chiral, there must be **four *different*** substituents attached. Such a carbon atom is called a *chiral center*.[2] Chirality may be illustrated by considering a series of substituted methanes.

Methane itself (www.bluffton.edu/~bergerd/Models/chiral2.html) is obviously *achiral* (not chiral). It is easy to see that methane can be superimposed on its mirror image; nevertheless, you may want to test this. The same is true for chloromethane.

$$H \quad\quad\quad\quad\quad\quad H \quad\quad C \quad\quad H$$
$$C \quad\quad\quad\quad\quad\quad H \quad C \quad\quad\quad \textbf{or}\quad CH_3Cl$$
$$H \quad H \quad\quad\quad\quad\quad\quad\quad Cl$$
$$H$$

Bromochloromethane, with *three* different substituents on carbon, may be more difficult to see; but if you build models of the two "different" molecules below you will find that they can be superimposed.

$$H \quad\quad H \quad\quad\quad\quad\quad\quad Br \quad\quad H$$
$$Br \quad C \quad\quad\quad\quad\quad\quad\quad H \quad C$$
$$Cl \quad\quad\quad\quad\quad\quad\quad\quad\quad Cl$$

However, a methane with *four* different substituents, such as bromochlorofluoromethane, is chiral (www.bluffton.edu/~bergerd/Models/chiral3.html). Build models of the two different molecules below. You will see that they are mirror images and cannot be superimposed! Such molecules are *enantiomers*[3] of each other.

---

[1] "Handedness."

[2] A chiral center need not be a carbon atom, as long as there are four different groups attached. For example, it is possible to have chiral ammonium ions or chiral silanes (compounds of silicon). In neutral nitrogen, a lone pair is formally able to serve as a fourth group but, because of the very low inversion barrier for amines, "chiral" neutral nitrogen compounds usually exist as racemic mixtures. Phosphines (phosphorus compounds analogous to amines) *can* be chiral because the inversion barrier at phosphorus is very high.

[3] Enantiomers are stereoisomers which *are* mirror images of each other.

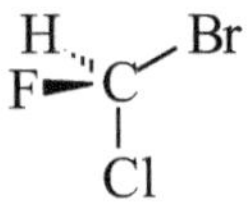

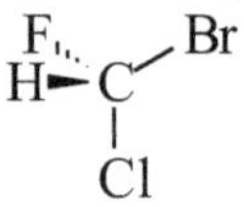

Use your models to explore the method explained in your text for determining whether configuration is R or S. Which of the two molecules above is R? Which is S?

Other molecules can be thought of as "substituted methanes." Those with four substituents – like 2-bromobutane and 2-butanol – are also chiral.

Build pairs of enantiomers for 2-butanol and 2-bromobutane. Draw the R and S configurations of each. Of the two molecules shown below, which is R and which is S?

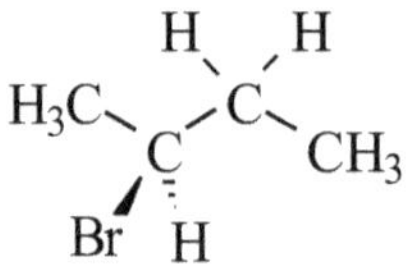

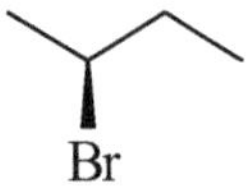

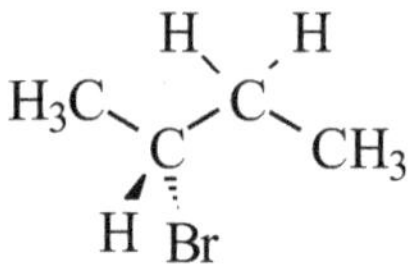

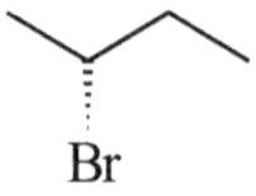

Molecules with more than one chiral center will obviously have more than two stereoisomers; in general, a molecule with n chiral centers will have 2^n stereoisomers. However, **this is a maximum** and is not always the case, as we will see.

For example, 3-bromo-2-butanol, with two chiral centers, will have $2^2 = 4$ stereoisomers. If each chiral center can be either R or S, obviously the stereoisomers will be RR, SS, RS and SR. These four are shown below; notice that each horizontal pair of isomers is a pair of enantiomers.

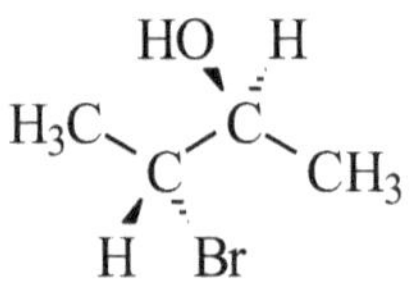

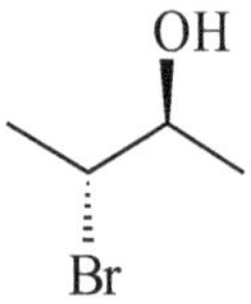

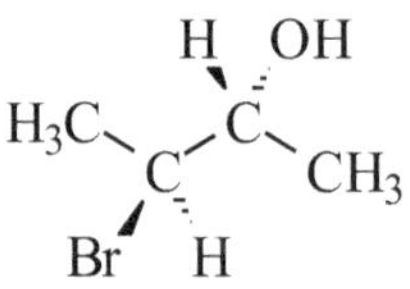

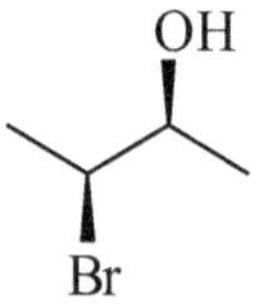

Assign the correct RS designation to each of the chiral centers in the molecules above, and name the molecules correctly.

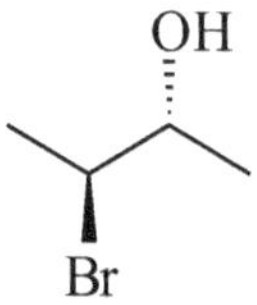

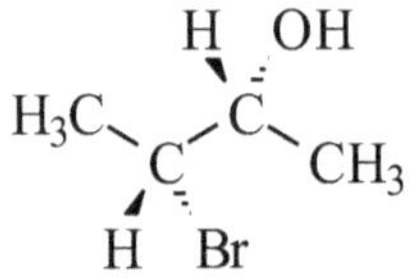

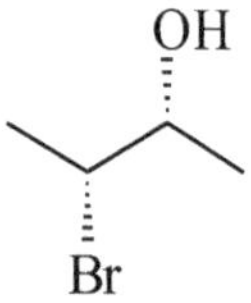

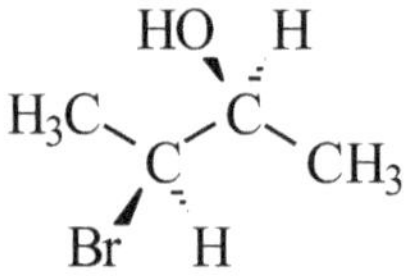

Remember that stereoisomers which are not mutual mirror images are called *diastereomers*. The top two molecules are diastereomers of the bottom ones.

When both chiral centers in a molecule have the **same substituents**, the molecule as a whole may or may not be chiral. If one half of a molecule is the mirror image of the other half, the molecule contains a plane of symmetry and **cannot** be chiral *even though it may contain chiral centers*.

Consider 2,3-dibromobutane. Like 3-bromo-2-butanol it has two chiral centers and therefore four (2^2) possible configurations: *RR, SS, RS* and *SR*. However, if you examine models of the four molecules below you will see that the bottom pair are identical! Molecules which contain chiral centers but are not themselves chiral are called *meso*, and we refer to them as (for example) *meso*-2,3-dibromobutane.

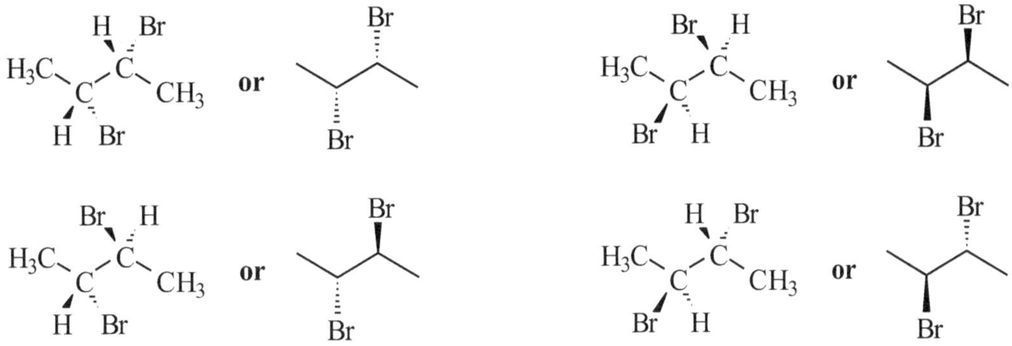

Assign R *or* S *configuration to each of the chiral centers in the molecules shown above. Which molecules contain a plane of symmetry?*

Fischer projections

www.bluffton.edu/~bergerd/Models/chiral6.html

Around 1890-1900, organic chemists were getting used to the idea that organic molecules are three-dimensional things, and that different arrangements of substituents in space give different molecules. But they were not at all used to thinking in three dimensions; in fact, the wedge-dash system of "perspective" drawing did not come into general use for about another 50 years.

Emil Fischer devised the system of Fischer projections, which allows a 3-D molecule to be correctly represented by a plane figure. There are two conventions associated with Fischer projections, but only the first is essential:

1. All vertical lines represent bonds going away from the viewer; all horizontal lines represent bonds coming toward the viewer.

2. The main chain of carbon atoms is laid out vertically by convention.

The two enantiomeric bromochlorofluoromethanes shown above can be represented thus:

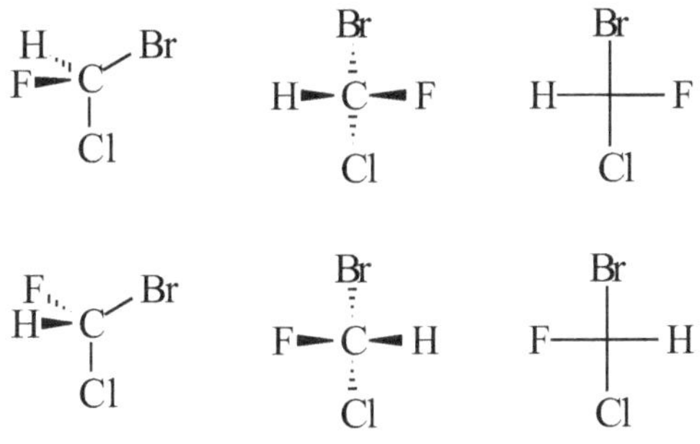

Satisfy yourself that the Fischer projections shown below correspond to the 2-bromobutanes, 3-bromo-2-butanols and 2,3-dibromobutanes shown in Section A.

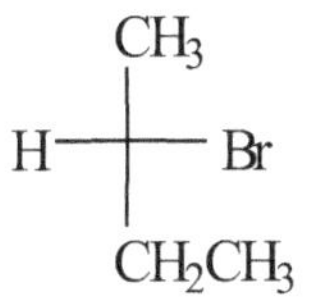

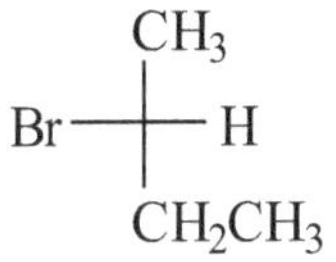

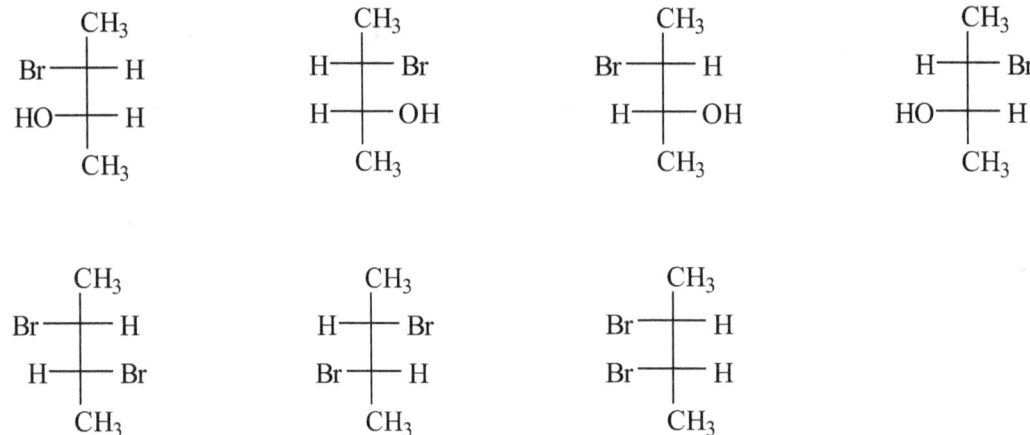

R,R-2,3-dibromobutane   S,S-2,3-dibromobutane   *meso*-2,3-dibromobutane

One advantage of a Fischer projection is that it becomes immediately obvious when a molecule is *meso*.

Rules for manipulating Fischer projections

www.bluffton.edu/~bergerd/Models/chiral7.html

The rules for manipulating Fischer projections as given in your text may be incomplete. The full rules are as follows.

1. An entire Fischer projection, as a single unit, may be rotated 180° only (**not** 90°), and *only in the plane of the paper*.

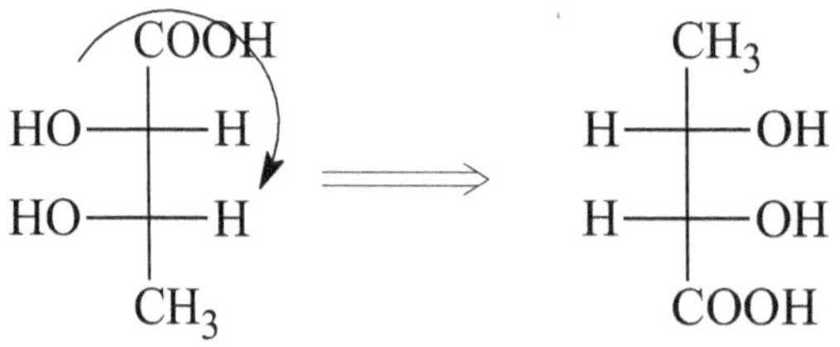

$$
\begin{array}{ccc}
& \text{COOH} & \\
\text{HO}\!\!-\!\!\!&\!\!\!-\!\!H & \\
\text{HO}\!\!-\!\!\!&\!\!\!-\!\!H & \\
& \text{CH}_3 &
\end{array}
\qquad\Longrightarrow\qquad
\begin{array}{ccc}
& \text{CH}_3 & \\
\text{H}\!\!-\!\!\!&\!\!\!-\!\!\text{OH} & \\
\text{H}\!\!-\!\!\!&\!\!\!-\!\!\text{OH} & \\
& \text{COOH} &
\end{array}
$$

2. Any **three** groups may be rotated, with the fourth remaining fixed, and the result will be a molecule identical to the starting molecule.[1] If we take the Fischer projection of one of our 2-bromobutanes and perform such a rotation, we can easily see that it is S-2-bromobutane. (Remember that, in the projection on the right, the H atom is pointing away from us.)

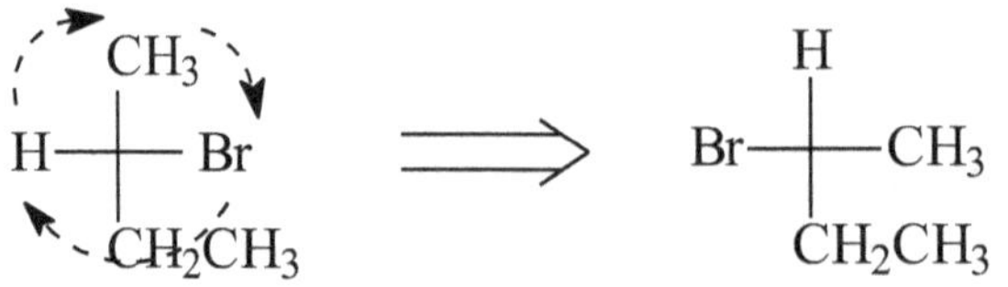

$$
\begin{array}{ccc}
& \text{CH}_3 & \\
\text{H}\!\!-\!\!\!&\!\!\!-\!\!\text{Br} & \\
& \text{CH}_2\text{CH}_3 &
\end{array}
\qquad\Longrightarrow\qquad
\begin{array}{ccc}
& \text{H} & \\
\text{Br}\!\!-\!\!\!&\!\!\!-\!\!\text{CH}_3 & \\
& \text{CH}_2\text{CH}_3 &
\end{array}
$$

Using your model kit or the online models, you should satisfy yourself that the pairs of Fischer projections above represent identical molecules.

---

[1] This is equivalent to rotating around a single bond.

Experimental Techniques

The following experiments are intended primarily as practice in necessary techniques in the organic chemistry laboratory. Most or all of these techniques will be new to you, but they will be used in other experiments in this laboratory manual.

The indispensable reference for these experiments is Zubrick's *Organic Chem Lab Survival Manual*.

Not all of these experiments will be used in any given academic year.

Synthesis of acetanilide

In this experiment, you will make *acetanilide* by reacting *acetic anhydride* with *aniline*. The reaction is as shown:

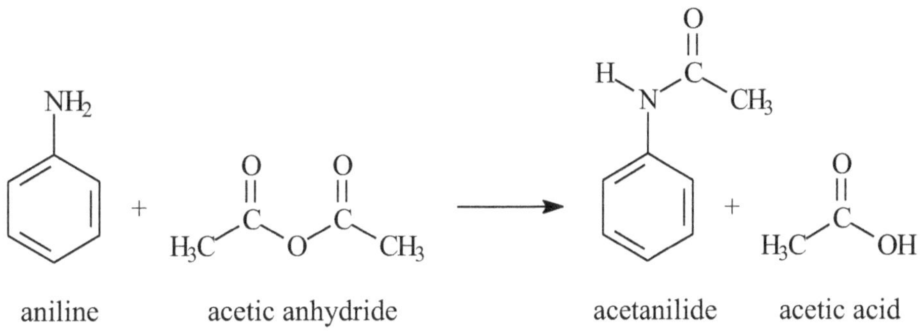

aniline   acetic anhydride   acetanilide   acetic acid

We have not yet discussed the functional groups in these molecules. However, the reaction itself is not our major concern in this experiment. You will gain experience in a useful technique for the purification of solid substances: *recrystallization*.

Determine the hazards of each of the chemicals shown by looking them up in the Merck Index or another suitable reference. Compare to the minimum safety standards for this experiment, given below.

You must hand in the pre-lab to be admitted to the laboratory.

Techniques used from Zubrick, 9th Ed: melting points (Chapter 12); recrystallization (Chapter 13)

Minimum Safety Standards for this experiment

1. Hot glass looks the same as cold glass! Before picking up a piece of glassware, be sure to check that it is cool enough to handle.

2. Reagents which have an odor or an appreciable vapor pressure may not be used outside the hood except in closed containers.

3. Look up the MSDS for each reagent used. More specific cautions and procedures are given below.

4. You will be boiling water during this experiment; be cautious and avoid steam burns.

5. Be sure to wash your hands after mixing the initial reaction solution to remove any traces of aniline and acetic anhydride. Wash after handling any other substance produced in this experiment. Washing after handling laboratory or household chemicals is always good procedure.

Disposal

acetanilide

Recrystallized product should be placed in the waste container provided after you obtain your melting point and final mass. Smaller amounts may be cleaned up with soap and water (not water alone!)

acetic anhydride

Residues should be cleaned with water and flushed down the sink.

aniline

Amounts larger than 1 mL should be cleaned with acetone or other solvent, and placed in a waste container in the hood. Residues should be cleaned with soap and water and flushed down the sink.

solid waste

May be disposed of in the waste-paper basket; residues in flasks may be cleaned with soap and water and flushed down the sink.

liquid filtration waste (contains acetic acid, acetanilide)

Clean with soap and water and flush down the sink.

Procedure

1. Put 9.1 mL aniline in a 500-ml Erlenmeyer flask. Add 60 mL water to the flask.

    Aniline is colorless when pure, but oxidizes slowly on exposure to atmospheric oxygen. The aniline you will use is brown due to this oxidation. The colored impurities will be removed during the purification of the product. It takes very little of a colored impurity to impart color to a substance.

2. Add 10 mL acetic anhydride to the aniline in 3 or 4 small portions. Swirl the flask while you add the acetic anhydride, and continue to swirl it for at least one minute. Describe any changes in the appearance of the material in the flask. **If you do not see any changes, consult the professor.**

3. Add 150 mL water. Heat the solution on a hot plate. Heat an additional 100 mL water in an Erlenmeyer flask on the same hot plate. Stir or swirl your product occasionally to help dissolve it.

4. When the solid is completely dissolved, remove the flask from the heat and add between 1 and 1.5 g activated carbon (charcoal).

    You may have a dark brown oil in the bottom of the flask; this oil contains some of your product and the process is called oiling out. Oiling out results from a solution temperature that is above the melting point of the impure product, without enough solvent to completely dissolve that product. Normally this would be a problem, but in this case it is OK. Most of the oil comes from high-molecular-weight oxidation products of the aniline

    CAUTION: If the solution is near its boiling point, the addition of the charcoal may cause violent boiling! (Why?) Add the charcoal **slowly**. Its purpose is to absorb many of the impurities, including most of the brown color.

5. Return the flask to the hot plate and continue heating until it boils. Watch carefully to keep it from boiling over!

6. The solution will now be filtered to remove the solids, through fan-folded filter paper[1] in a funnel. The funnel must be kept hot to prevent the product from crystallizing and clogging the funnel.

    Put a wide-mouthed funnel in a ring stand and fold a filter paper to fit it, as discussed in Zubrick ("the famous fan-folded filter paper"). Place a clean 500-mL Erlenmeyer flask under the funnel. Notice that the flask sits on the hot plate! Suspend the funnel in the mouth of the flask **so that a small space is left all around**. This allows steam to warm the funnel as it escapes.

7. **IMMEDIATELY** pour half of the hot *water* through the filter; if you delay, the flask may crack. Now pour the hot *solution* through the filter.[2] Rinse the reaction flask with the remaining hot water, and pour the rinse through the filter.

    CAUTION: Be sure to work quickly so everything stays hot, but take care not to spill or to burn yourself.

8. If crystals have formed in the filtered solution, heat the solution on a hotplate to redissolve the crystals. Be sure to use one or two glass stirring rods to help prevent *bumping*.

    Bumping is sudden, violent boiling of a solution and may result in boilover or even knock over a flask! How do the stirring rods help prevent this?

9. When all of the crystals have dissolved, remove the stirring rods and place the flask on the bench top to cool slowly. Crystals will again form. Let the solution stand and cool for at least 30 minutes, until the flask is cool to the touch.

    The process of allowing crystals to grow slowly from a supersaturated solution is called *recrystallization*. It normally produces very clean product. As the crystals grow, only the similar molecules will attach to the crystal and all the impurities will remain in the solution. The slower the crystals grow, the more pure the product. Rapid crystallization will often trap impurities in the crystal.

    Why is water a good solvent for the recrystallization of acetanilide? *HINT*: consult the *Merck Index* to find the solubility of acetanilide in water.

10. Recover the crystals by vacuum filtration through a Buchner funnel. Rinse the crystals with two or three portions of about 30 mL of **cold** water. (Why cold?) Draw air over the crystals for several minutes to dry them. Transfer the product to a tared,[3] shallow dish and place the dish in the drying cabinet. Be sure the dish is properly labeled. Allow the crystals to dry, uncovered, for a few days until the crystals weigh the same on two consecutive days. This is called *drying to constant weight*.

11. When the crystals are dry, determine the mass and melting point of the product.

---

[1] See the discussion in Zubrick, at the end of the chapter on recrystallization.

[2] Take care not to allow the mixture in the funnel to rise higher than or flow over the filter paper! If you get charcoal in your filtrate you will need to hot-filter it again.

[3] That is, pre-weighed. Often containers are *tared* so that one can tell how much is in them: weigh the whole thing, then subtract the mass of the container!

For the report

Report your yield of acetanilide and its melting point. Compare to the accepted value of the melting point and account for any significant disparity.

Was the entire product recovered? If not, what happened to the rest? How could you improve the recovery?

Pre-laboratory assignment

1. Use the Merck Index to find the solubility of acetanilide in alcohol, and compare to its solubility in water. Account for the difference.

2. Use the Merck Index to find the solubility of acetanilide in hot water and in cold water. Why is it important to keep aqueous solutions of acetanilide hot?

3. Are there any extraordinary hazards present in this experiment? Defend your answer. Consider not only the chemicals used, but also their amounts and the conditions.

A Solvent-Free Reaction: the Aldol Condensation

Based on Doxsee, K.M.; Hutchison, J.E. *Green Organic Chemistry: Strategies, Tools and Laboratory Experiments*, Brooks-Cole (2004), pp. 115-119. Used with permission.

The aldol condensation is a powerful, general method for forming carbon-carbon bonds between two carbonyl compounds. The reaction is base-catalyzed and begins by deprotonation *alpha* (adjacent) to a carbonyl group. This generates a resonance-stabilized *enolate* anion, which is a nucleophile that is able to add to the carbonyl group on another reactant molecule. The product, an *aldol* – also called a *beta-hydroxycarbonyl compound* – can easily be dehydrated (lose water) to form a new carbon-carbon double bond that is conjugated with the remaining carbonyl group.

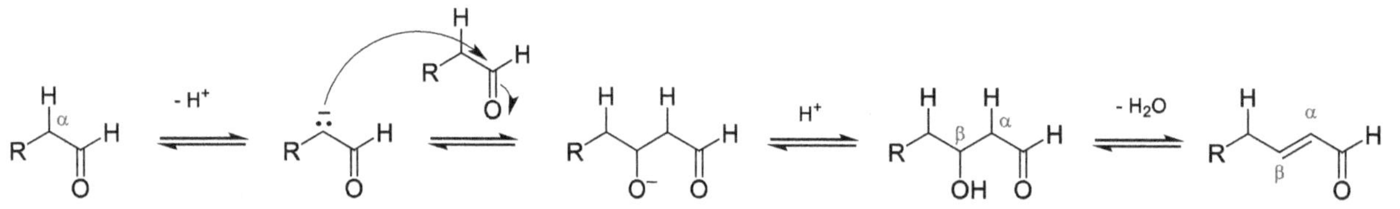

Alpha (α) and beta (β) positions are indicated in this mechanism.

If both compounds used were able to form enolates, the result would be complex mixture of products. Typically, to avoid this, pairs of compounds are chosen in which only one has *alpha* hydrogen atoms, so that only one can be the nucleophile.

In this experiment we will use 1-indanone as the nucleophile and 3,4-dimethoxybenzaldehyde as the aldol acceptor. Note that only one compound of this pair has α-hydrogen atoms.

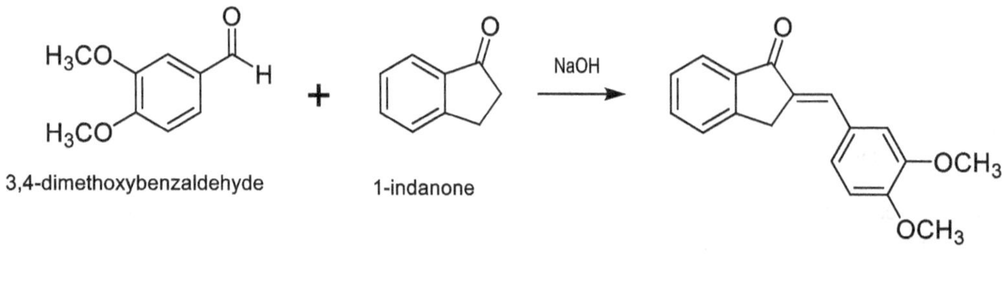

3,4-dimethoxybenzaldehyde   1-indanone

This reaction is called "solvent-free," but actually occurs in the liquid phase because of *melting-point lowering* caused by mixing, so that the reactants form a liquid mixture before they react.

Melting points of impure substances are lower and broader than the melting points of pure substances. This phenomenon is one of the points of this experiment and is clearly demonstrated while you are carrying out the reaction: as you mix the two solid reactants, they dissolve in each other to form a liquid.

This is also a recrystallization experiment: we are using the technique called *one-solvent* recrystallization, not because we are using a pure solvent (we aren't) but because we don't change the solvent composition in order to coax our product to crystallize. One-solvent recrystallizations require a solvent that will dissolve a lot more material when hot than it will when it is cold.

Techniques used from Zubrick, 9ᵗʰ Ed: Melting point determination and mixed melting points (Chapter 12); recrystallization and vacuum filtration (Chapter 13).

Minimum Safety Standards for This Experiment

1. 1-Indanone is an irritant. Containers should be kept closed when not in use. Wash surfaces that it has been spilled on.

2. 3,4-Dimethoxybenzaldehyde is an irritant. Containers should be kept closed when not in use. Wash surfaces that it has been spilled on.

3. Sodium hydroxide is a strong caustic, and the dust is hazardous to eyes and mucous membranes. Use gloves and eye protection when grinding and wash thoroughly after handling. Wipe up spills with plenty of water.

4. 10% Hydrochloric acid is corrosive. Handle with care and wipe up spills with plenty of water.

5. 90% ethanol is a strong irritant, particularly to eyes and mucous membranes. It should be kept in the hood.

6. Your product, 2-(3,4-dimethoxyphenylmethylene)-1-indanone, has unknown toxicity and should be handled with appropriate care. Wash surfaces that it has been spilled on.

Disposal

1-indanone	Spills should be swept up and thrown in the trash. Residues should be cleaned with soap and water.
3,4-dimethoxy-benzaldehyde	Spills should be swept up and thrown in the trash. Residues should be cleaned with soap and water.
sodium hydroxide	Spills should be swept up and flushed down the sink. Residues should be cleaned with plenty of water.
10% hydrochloric acid	Spills should be cleaned with plenty of water. Larger quantities should be flushed down the sink.
90% ethanol	Spills should be wiped up; allow the paper towels to dry in the hood before discarding. Leftover liquid in vessels or graduated cylinders should be flushed down the sink.

mother liquor	The liquid remaining after recrystallization should be placed in the waste bottle provided.
Product	Spills should be swept up and thrown in the trash. Larger quantities (more than 0.25 g) should be placed in the waste bottle provided.

Procedure

If powdered sodium hydroxide comes in contact with your skin, you should be sure to rinse thoroughly in running water within a few minutes.

1. Place 0.25 g of 3,4-dimethoxybenzaldehyde and 0.20 g of 1-indanone in a test tube. Crush and scrape the two solids together with a spatula until they form an oil, which will be colored yellow or brown. Do not break the test tube!

2. Sharing with other students, grind about 0.1 g of sodium hydroxide finely, using a mortar and pestle. Add roughly 50 mg of the ground sodium hydroxide to your reaction mixture. Continue mixing and scraping until the mixture becomes solid.

3. Allow the mixture to stand for 15 minutes, then add about 2 mL of 10% HCl solution. Scrape well to dislodge the product mixture from the walls of the test tube. Check the pH of the solution to make sure it is acidic; if it is not, add a little more HCl solution, mix and test again.

4. Isolate the product by vacuum filtration. Pull air through the solid for at least 10 minutes, to facilitate drying. Determine the mass of your crude product.

5. Recrystallize the product from 90% ethanol/10% water, using some of your solvent to rinse the remaining product from your test tube. You should not require more than 20 mL of solvent for this recrystallization.

6. Allow your product to dry for at least an hour before determining its mass and melting point. A typical melting point for this product is 178-181°C.

For the report

The experimental section for this report should be brief and to the point. Don't forget to include gram amounts in the experimental section, and **only** there.

Report the atom economy of this experiment, and the yield (as a percentage) that you obtained both before and after recrystallization. Is this experiment wasteful? Why or why not?

Pre-laboratory assignment

1. Determine the mole ratio of 3,4-dimethoxybenzadehyde and 1-indanone that you will use.

2. Calculate the atom economy for this reaction, by comparing the theoretical yield of product to the amounts of starting materials you expect to use. Since sodium hydroxide is a catalyst, it is not counted in this calculation! (See the Atom Economy handout for more.)

3. What is the purpose of hydrochloric acid during workup?

4. Why is it important to use minimal amounts of hot solvent during recrystallization?

5. Are there any extraordinary hazards present in this experiment? Defend your answer. Consider not only the chemicals used, but also their amounts and the conditions.

Extraction and Isolation of Acetylsalicylic Acid[1]

Extraction from a solid by a liquid is a technique that goes back two thousand years in the laboratory, and considerably further in the kitchen: tea, coffee and soup are all made using this technique.

In this experiment we will extract the active ingredient from aspirin tablets, using an organic solvent. You will determine a yield based on the amount of aspirin per tablet, listed on the label of the aspirin we use.

Because aspirin dissolves well in alcohols, we will soak the material to be extracted in warm solvent, filter to remove the solid residue, and crystallize the extracted material.

Techniques used from Zubrick, 9th Ed: melting points (Chapter 12); mixed-solvent recrystallization and gravity filtration (Chapter 13)

You must hand in the pre-lab to be admitted to the laboratory.

Minimum Safety Standards for this experiment

1. Hot glass looks the same as cold glass! Before picking up a piece of glassware, be sure to check that it is cool enough to handle.

2. Reagents which have an odor or an appreciable vapor pressure may not be used outside the hood except in closed containers.

3. Look up the MSDS for each reagent used. More specific cautions and procedures are given below.

Disposal

All liquids used or produced in this experiment may be flushed down the sink.

All solids used or produced in this experiment may be placed in the wastebasket, <u>after</u> allowing any liquid to evaporate.

Procedure

1. Place twelve aspirin into a flask; record the number of aspirin tablets you use and the amount of aspirin per tablet. Add about 50 mL of 70% isopropyl alcohol ("rubbing alcohol"). Heat the flask until the solvent is boiling; while heating, begin to crush the aspirin tablets in the flask. Remove the flask from the heat *as soon as it boils*.

2. Using a heavy stirring rod, finish crushing the aspirin tablets in the warm alcohol. Remove the stirring rod and allow the mixture to stand for fifteen to twenty minutes, to allow the active ingredient to dissolve.

---

[1] Based on a procedure from Cobb, C.; Fetterolf, M.L. *The Joy of Chemistry: The Amazing Science of Familiar Things*. Amherst, NY: Prometheus Books (2010).

3. Filter the mixture through a coffee filter in a wide-mouthed funnel, into a 250-mL flask. Add 100-150 mL of cold tap water to the flask, and allow the mixture to stand for several minutes. If you don't have crystals after 20 minutes, place your mixture in an ice bath.

4. Vacuum filter your crystals of acetylsalicylic acid in a Buchner funnel, and run air through the funnel for ten to fifteen minutes to dry the crystals. Transfer your product to a labeled evaporation dish and allow it to stand for at least 48 hours in the drying cabinet before weighing it, so that the water may evaporate.

5. Obtain a melting point and compare it to the expected melting point for acetylsalicylic acid.

For the report

Based on the number of aspirin tablets you used and the amount of aspirin per tablet, how much acetylsalicylic acid did you expect to produce? How much did you actually obtain? What was your yield?

Prelaboratory questions

1. Draw the structure of acetylsalicylic acid.

2. Look up the solubility of acetylsalicylic acid in the *Merck Index*. Based on its structure, explain the difference in its solubility in alcohol versus water.

3. Why do we use warm solvent for the extraction, rather than cold or room-temperature solvent?

4. Suppose you use 15 @ 300 mg aspirin tablets.

    a. What is the total amount of acetylsalicylic acid contained in the tablets you used?

    b. Suppose you obtained 1.8 grams of product. What is the relative yield?

5. Are there any extraordinary hazards present in this experiment? Defend your answer. Consider not only the chemicals used, but also their amounts and the conditions.

Isolation of caffeine from coffee or tea[1]

In this experiment, you will isolate a naturally occurring organic chemical from one of its natural sources. In so doing, you will learn some common purification techniques.

Experiments of this type were the first stirrings of organic chemistry, and, indeed, 180 years ago scientists used "organic chemistry" to mean what we think of as "biochemistry" today: the chemistry of living things. The first organic chemists were concerned with the isolation and identification of substances from materials that were once living, just as you will do in this experiment.

---

[1] Based upon "Caffeine Extraction from Tea – A Simplified Procedure," by Edward G. Neeland, Okanagan University College, Kelowna, British Columbia, Canada.

As in all laboratory procedures, you should look up any known hazards associated with the substances you will be using or preparing. You may also want to look up the structure of caffeine and think about why this procedure works, on the basis of that structure.

Techniques used from Zubrick, 9th Ed: extraction and washing (Zubrick Chapters 15, 35), evaporation on a steam bath (Zubrick chapters 17), drying an organic liquid (Zubrick Chapter 10), mixed-solvent recrystallization and vacuum filtration (Zubrick Chapter 13)

You must hand in the pre-lab to be admitted to the laboratory.

Minimum Safety Standards for this experiment

1. Hot glass looks the same as cold glass! Before picking up a piece of glassware, be sure to check that it is cool enough to handle. In particular, mind the caution about allowing dichloromethane to contact hot water.

2. Reagents which have an odor or an appreciable vapor pressure may not be used outside the hood except in closed containers.

3. Look up the MSDS for each reagent used. More specific cautions and procedures are given below.

4. 6M sodium hydroxide[1] is a strong caustic. Treat it appropriately.

5. You should be careful to avoid spillage during the extraction portion of the experiment. Not only will you spoil your results, but–because of the hazard associated with dichloromethane–you should avoid contamination.

6. You will be recrystallizing from a mixture of 2-propanol (a purer form of "rubbing alcohol") and mixed hexanes, which are hydrocarbons. These solvents present a flammability hazard, and inhalation of large quantities can cause lightheadedness; use them in the hood as far as possible.

7. Wash your hands after completing the experiment.

Disposal

caffeine	Discard in the wastebasket.
dichloromethane	The dichloromethane you use will be evaporated. Wipe up spillage with paper towels and allow the dichloromethane to evaporate in the hood. The paper towels may then be disposed of in the wastebasket.
drying agents	Allow to stand in the hood until dry; then throw in the wastebasket.
hexanes	Recrystallization solvents must be placed in the waste bottle provided.
2-propanol	Recrystallization solvents must be placed in the waste bottle provided. Pure 2-propanol may be flushed down the sink.
sodium hydroxide solutions	Flush down the sink.

---

[1] What does "6M" (stands for "six molar" or "six moles per liter") mean? If you don't know, consult your general chemistry textbook!

tea bags	Discard in the wastebasket.
tea residue, including aqueous waste from the extraction	Clean with soap and water and flush down the sink.

Procedure

In this experiment you will use clamps and rings to hold your glassware, something you will continue to do throughout the year. Unlike general chemistry, things will not be set up for you! Use your common sense: if something needs to be supported, support it. Use clamps to hold things by the neck, not around the body. Rings are better for resting larger, globular objects like separatory funnels.

You may need to manipulate hot glassware with tongs. I suggest that you practice using the tongs before you spill something!

1. Bring 100-120 mL of tap water to a boil in a beaker. While it is heating, get between six and seven grams of either coffee or tea in a coffee filter. You will be instructed which to use.

2. When the water is boiling, remove it from the heat and add your coffee or tea to the water. Let it steep for about five minutes, stirring occasionally.

3. While your coffee or tea is steeping, place your coffee filter in the filter funnel provided to you. Filter the coffee or tea when it has finished steeping. Cool the coffee or tea to below 35°C, using an ice bath.

4. Extract the coffee or tea three times with 20-25 mL portions of dichloromethane. Wash the combined organic layers twice with 20 mL portions of 6M NaOH, and once with 20 mL of tap water. Remove the organic layer and dry it over the drying agent provided (calcium chloride).

5. Evaporate the dried organic layer in a beaker, using a steam bath. Recover as much product as you can from the beaker after the solvent has evaporated. Weigh your product. Obtain a melting point of your product, comparing it to an authentic sample of caffeine.

    Although pure caffeine is white, the residue you obtain may be green or brown due to the presence of chlorophyll and tannins. Determine the mass of the crude product and find its melting point.[1] How does the melting point compare with the literature value for caffeine?

6. Report your results (mass of coffee or tea, mass of caffeine obtained) to the instructor so that they may be compiled with those of other students. Compare your results to those obtained by your lab mates in your report.

7. **If there is time**, purify the crude caffeine by *recrystallization*.

    a. Place your crude caffeine[2] into a 25-mL Erlenmeyer flask. Heat a few mL of 2-propanol until it is almost boiling; then dissolve your caffeine in a *minimum* amount of hot 2-propanol. You will want to keep your flask warm in a hot water bath.

---

[1] It is not necessary to do this right away; you can set aside a sample of your crude caffeine in a melting-point capillary tube for later.

[2] If you do not have at least 80 mg of crude caffeine, you should combine your crude caffeine with another student's.

b. Now allow your solution to cool to room temperature. With careful stirring, add 1 mL of hexanes and allow the mixture to stand for a few minutes. Collect the crystals of caffeine by vacuum filtration in a Hirsch funnel, and wash with *small* amounts of hexanes.

c. Determine the mass and melting point of the purified caffeine. Did you lose much in the recrystallization step?

For the report…

The report will have a cover page with an abstract, and a body. No experimental section is required. The report should answer the questions asked in the procedure and pre-lab, with any additional comments you may wish to make about the experiment. All sources of information will be properly referenced. The body of the report will be a unified narrative, with a beginning, middle and end.

Determine the weight percentage of caffeine in your tea or coffee sample.[1] Suggest an extraction method that might improve your yield of crude caffeine.

Prelaboratory questions

1. You will extract the tea three times with dichloromethane. Why not six, or ten times? Why not just once? (HINT: see Zubrick, Chapter 37, "Theory of Extraction"!)

2. You will wash the dichloromethane solution with strong base. Explain a possible reason for this. (No, dichloromethane is not an acid. If that were so, it would be hazardous to treat it with concentrated base! Why? What is the extraction doing for you?)

3. Are there any extraordinary hazards present in this experiment? Defend your answer. Consider not only the chemicals used, but also their amounts and the conditions.

Biosynthesis of ethanol by fermentation and distillation

Introduction

Sugar cane is the most useful source of ethanol for biofuels, because of its high sugar content. Molasses (the byproduct of sugar refining) is typically processed into alcohol rather than using the refined sugar itself, which is more valuable. Alcohol is obtained from molasses by fermentation of its sugars by yeast, and is isolated by distillation.

Other crops that contain sugar can be used to make ethanol; corn, for example, or sugar beets. But the yield of alcohol, and therefore the efficiency of the process, depends on sugar content, because sugar is what yeast converts into ethanol. Energy yield estimates indicate that sugar cane yields ten times the energy required to process it into ethanol, while corn yields very little, if any, energy profit.

Fermentation vats normally have "fermentation locks," U-tubes with water filling the curve. These serve to permit the exit of carbon dioxide (a by-product of fermentation) and prevent the entrance of air. Oxygen in the air would further oxidize ethanol to acetic acid.

In this experiment we will use white sugar, to avoid the sludge that is left behind from molasses fermentation.

---

[1] The approximate amount of caffeine in black tea is 25-110 mg per teabag.

Techniques used, from Zubrick, 9th Ed: gravity filtration (Chapter 15[1]); distillation (Chapter 19, Class 1 and Class 3 – see Fig. 19.10)

Minimum Safety Standards for this experiment

1. Hot glass looks the same as cold glass! Before picking up a piece of glassware, be sure to check that it is cool enough to handle.

2. The ethanol we will produce is no more toxic than any other distilled liquor, but it is not likely to taste very good, and it's being generated in glassware that is not cleaned to food-quality standards. So don't try to drink it.

Disposal

All wastes and products produced in this experiment may be thrown in the trash or flushed down the sink, with one exception: **Glass waste is to be placed in the glass disposal container provided.** If it is hot, sure to let it cool before disposing of it! The disposal container is plastic-lined.

Procedure

In this experiment you will use clamps and rings to hold your glassware, something you will continue to do throughout the year. Unlike general chemistry, things will not be set up for you! Use your common sense: if something needs to be supported, support it. Use clamps to hold vessels by the neck, not around the body. Rings are better for resting larger objects like heating mantles.

We will concentrate on producing ethanol with high purity. We will not calculate "yield," because there is no good way of knowing what the yield is supposed to be. For that reason, you don't need to be terribly precise about your measurements in steps 1 and 2.

1. Use glycerine as lubricant to insert the short end of a bent glass tube into a one-hole rubber stopper that fits the top of a 250-mL Erlenmeyer flask.

2. Dissolve 25-30 g sugar in about 100 mL of water in a 250-mL flask, warming if necessary.

3. Cool the flask to 30° or below.

4. Add 0.5-1 g baker's yeast and stir well. Label your flask properly.

5. Fill a test tube about half-full with water. Stopper your flask securely with the stopper through which you placed the bent glass tube, and place the other end of the tubing in the water. This is a fermentation lock: carbon dioxide can get out, but air cannot get back into the fermentation flask. Allow your flask to stand for at least a week, checking periodically to ensure that the water in the test tube is not getting too low.

6. Open your flask and gravity-filter your fermentation mixture into your 250-mL round-bottomed flask; this will take a while. Set up a simple distillation, being sure to leave enough room under your still pot for a ring support for a heating mantle. Be sure to clamp your apparatus properly.

7. After the instructor approves your still setup, distill your mixture rapidly until you have 25-50 mL of distillate.

---

[1] Remember that the description in Zubrick is for hot filtration; we are not doing that, we are merely separating yeast from liquid fermentation products.

8. Set up a fractional distillation with your initial distillate in a clean still pot, and heat carefully; try to keep the distillation rate at or below a drop every five seconds. Record the temperature range for each fraction collected, stopping the distillation at or below 95-97°C. Fractions are identified by a rapid change in temperature and a concurrent increase/decrease in distillation product (see Zubrick). Try to keep your fractions to a small temperature range, not wider than about 5°.

   Note: Normally there is not time to collect more than a single fraction, which will be your purest ethanol anyway.

9. Determine the density of each fraction by weighing a portion that is precisely measured using a graduated cylinder. Use the table below to determine the alcohol content in each fraction.

10. After determining its density, dispose of your product by flushing it down the sink.

Aqueous Ethanol (EtOH) content							
Density g/mL	% EtOH by weight	% EtOH by volume	g EtOH per 100mL	Density g/mL	% EtOH by weight	% EtOH by volume	g EtOH per 100mL
0.989	5	6.27	4.95	0.856	75	81.30	64.17
0.982	10	12.44	9.82	0.843	80	85.49	67.48
0.975	15	18.54	14.63	0.831	85	89.48	70.63
0.969	20	24.54	19.37	0.828	86	90.25	71.23
0.962	25	30.46	24.04	0.826	87	91.02	71.84
0.954	30	36.25	28.61	0.823	88	91.77	72.43
0.945	35	41.90	33.07	0.821	89	92.53	73.03
0.935	40	47.40	37.41	0.818	90	93.27	73.62
0.925	45	52.72	41.61	0.815	91	93.99	74.19
0.914	50	57.89	45.69	0.813	92	94.72	74.76
0.903	55	62.89	49.64	0.810	93	95.44	75.32
0.891	60	67.74	53.47	0.807	94	96.11	75.86
0.880	65	72.43	57.17	0.804	95	96.79	76.40
0.868	70	76.95	60.74	0.789	100	100.00	78.9

For the report

Report the volume and ethanol content of your distillate(s). Calculate how much total ethanol you have made, excluding any remaining water.

Extra credit: Investigate the different types of yeast used for ethanol production, and their efficiency compared to bakers' yeast.

Pre-laboratory questions

1. What component of molasses is processed into ethanol by yeast?

2. Why do we need to allow gas to escape from our fermentation vessel?

3. Why do we need to prevent air from entering the fermentation vessel?

4. Why does a distillation apparatus need an opening in it, somewhere?

Resolution of a Racemic Mixture: α-Methylbenzylamine

Since enantiomers have identical physical properties, it can be very difficult to resolve a racemate (a 50:50 mixture of enantiomers) into the two enantiomers of which it is composed. The only way this can be accomplished is by use of another chiral compound.[1]

In this experiment, you will use optically pure (+)-tartaric acid to resolve a racemic mixture of α-methylbenzylamine or (±)-MBA. As a Brønsted base, MBA will react with tartaric acid to form a salt. Half of the salt will be (+)-ammonium-(+)-tartrate and the other half will be (-)-ammonium-(+)-tartrate. These two compounds are chiral, but they are not enantiomers of each other. Rather, they are diastereomers. Diastereomers are not mirror images of each other, and do **not** have identical physical properties. It happens that one of the diastereomeric ammonium tartrates is less soluble in methanol than the other, so the salts are separable by selective crystallization.

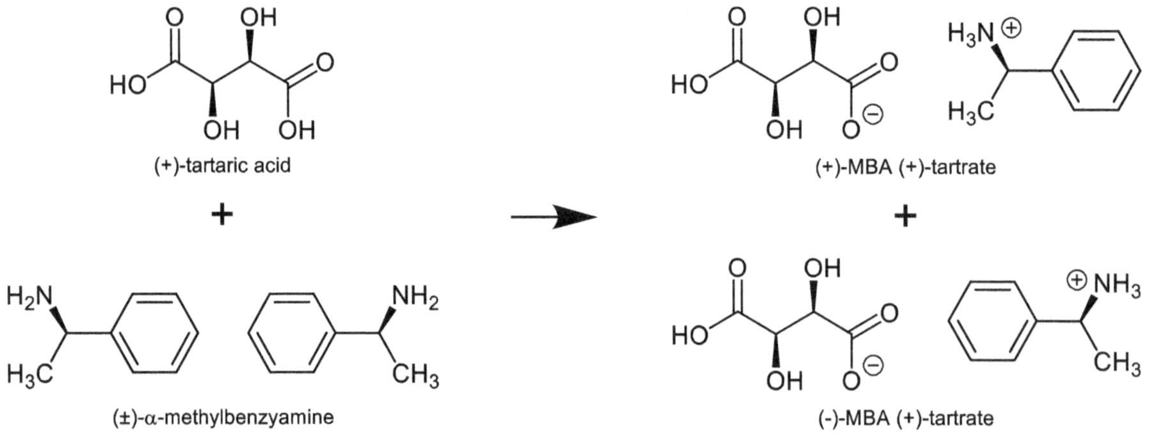

(+)-tartaric acid

+

(±)-α-methylbenzyamine

(+)-MBA (+)-tartrate

+

(-)-MBA (+)-tartrate

What sort of reaction is shown above?[2] Classifying the type of reaction will be helpful to you as you think about the purposes of the reagents used in this experiment.

---

[1] See Karty, section 5.11 for a discussion; see also the problem at the end of that section.

[2] See Chapter 6 of Karty!

Techniques used from Zubrick, 8ᵗʰ Ed: crystallization and vacuum filtration (Chapter 13), extraction (Chapters 15, 35), evaporation under reduced pressure (Chapter 21), polarimetry (Karty, Chapter 5).

Minimum Safety Standards for this experiment

1. Hot glass looks the same as cold glass! Before picking up a piece of glassware, be sure to check that it is cool enough to handle. In particular, mind the caution about allowing dichloromethane to contact hot water.

2. Reagents which have an odor or an appreciable vapor pressure may not be used outside the hood except in closed containers.

3. Look up the MSDS for each reagent used. More specific cautions and procedures are given below.

4. Tartaric acid is mildly corrosive and will irritate your eyes and mucous membranes if you touch them with tartaric acid on your fingers. Treat tartaric acid with respect and wash your hands thoroughly after handling.

5. α-Methylbenzylamine is smelly, caustic and of unknown toxicity. Treat it with respect, handle *only* in the hood, and wash your hands thoroughly with the dilute acid solution after handling. Wipe up ALL spills *immediately* and keep the paper towels in the hood; they MUST **NOT** BE THROWN IN THE TRASH before treatment with dilute acid as described in the **Disposal** section below.

6. You will use a dilute (ca. 0.1 mol/L) solution of hydrochloric acid to clean and deodorize α-methylbenzylamine residues. This solution is innocuous provided you take care to avoid contact with your eyes or mucous membranes, and wash your hands immediately after use.

7. You will be using a 2M solution of potassium hydroxide during this experiment. KOH is a strong caustic, and a 2M solution is rather concentrated. Treat it with respect, and wash your hands afterwards. **Be sure to remove jewelry when washing, so that you do not trap any of this reagent next to your skin**.

8. Be sure to wash your hands (including *rinsing with dilute acid*) after mixing the initial reaction solution to remove any traces of the reagents. Also wash after handling any other substance produced in this experiment.

Disposal

amine/tartaric acid salt	Small residues may be cleaned with soap and water and flushed down the sink. Be sure to rinse the newly cleaned item with dilute acid.
dichloromethane	Spills may be allowed to evaporate if inside the hood. You will be evaporating all the dichloromethane you use, so there is no need to dispose of any.
methanol solution of *α-methylbenzylamine*	Place in the waste bottle provided; the instructor will treat it with acid to deactivate the amine.
methanol	May be flushed down the sink.
mother liquor from crystallization	May be flushed down the sink; be sure to rinse the container with dilute acid.
α-methylbenzylamine	Residues should be cleaned with the dilute acid solution provided and flushed down the sink. **Do not try to clean with acetone or water; you must use acid to remove the amine odor. All objects**

contaminated with this substance must be treated with LARGE quantities of the dilute acid, until the odor has vanished.

potassium hydroxide solution	May be flushed down the sink.
residual aqueous solution from extraction	Flush down the sink **in the hood.** Wash glassware with water, then with **copious** quantities of acid, then with water again.
tartaric acid	Residues should be cleaned with water and flushed down the sink.

All glassware, bench tops and trash cans WILL be checked for odor. **Failure to properly treat waste is a safety violation.** If the person responsible cannot be determined, the entire section will be penalized.

Procedure, First Week

1. In a 500-mL Erlenmeyer flask, mix 18.0 g (+)-tartaric acid with 260 mL methanol. Heat the mixture on a hot plate until the tartaric acid dissolves, then remove from heat. **Carefully** add 14.5 g (±)-MBA.[1] Keep the amounts used within three or four percent of the amounts called for!

   CAUTION: Exothermic reaction! Boilover hazard!

2. Let the solution cool for about 15 minutes, stopper the flask, and let the solution stand for a week to allow crystal formation. Be sure to label your flask correctly. After the lab period is over, the professor may add a seed crystal to your solution to encourage it to form the "correct" crystals.[2]

Pre-laboratory questions, week 1

1. Why is it impossible to just distill the (+)-α-methylbenzylamine away from the (-) enantiomer?

2. What does making the tartrate salt do for you? Why can't you just use acetic acid?

3. Why is dilute acid so effective at destroying the odor of the amine? How does it help in washing the amine away? (HINT: what chemical reaction is involved? What does the reaction do for the water solubility of the amine?

4. Are there any extraordinary hazards present in this experiment? Defend your answer. Consider not only the chemicals used, but also their amounts and the conditions.

Procedure, Second Week

1. When crystals have been obtained, recover them by vacuum filtration, washing them with a small amount of methanol.

   Weigh your crystals and report the yield, based on the total amount of MBA used. What yield would you expect if your separation were perfect, that is, if you got 100% of a single diastereomer?

2. Dissolve the crystals in 70 mL of 2M potassium hydroxide. (What is the purpose of the potassium hydroxide?) **Be sure all your crystals have dissolved!**[1] Extract this mixture twice with 20-mL

---

[1] Measure the MBA by <u>volume,</u> in the hood; use its density to determine the mass.

[2] You may get large, prism-like, crystals (a single diastereomer) or small needle-like crystals (contaminated with the other diastereomer). You can still get an optical activity reading with the needle-like crystals.

portions of dichloromethane, using the separatory funnel. Combine the dichloromethane extracts in a 100-mL round-bottomed flask, being careful to avoid visible water inclusions.

WARNING. From this point on, your product will be **smelly**. It must be either contained or kept in the hood *at all times*. Thoroughly rinse anything your product touches with the acid wash solution provided.

3. Evaporate the dichloromethane under reduced pressure on the rotary evaporator, using a warm water bath. **Be sure to weigh your product when it is free of dichloromethane** (use a stoppered flask for this; you may use your round-bottomed flask with the ground glass stopper, but be sure to set it in a beaker so it can't tip over).

4. Dilute your product to a **total** volume of between 10 and 12 mL with methanol and record the exact volume after dilution. Calculate the concentration in grams of MBA per milliliter of solution. Mix well with a stirring rod.

5. Use the polarimeter to examine your MBA solution. **The specific rotation of MBA is 40.3°.**[2] Determine whether you have made (+)-MBA or (-)-MBA, and record the optical rotation you observe. (From this you can determine the optical purity of your product.) When you have taken this measurement, put the MBA solution into the appropriate waste bottle.

Pre-laboratory questions, week 2

1. What is the identity of the crystalline product that you begin with this week? Where did it come from, and what chemical reaction produced it?

2. What is the purpose of treating the crystalline product with strong base? (HINT: think about the chemistry involved in making the crystals. How will the base undo that chemistry?)

3. When you perform the extraction, which layer do you expect to be on the bottom? Why?

4. Why is it reasonable to evaporate dichloromethane under reduced pressure, without losing an appreciable quantity of α-methylbenzylamine? (HINT: what are the boiling points of dichloromethane and α-methylbenzylamine?)

5. Are there any extraordinary hazards present in this experiment? Defend your answer. Consider not only the chemicals used, but also their amounts and the conditions.

---

[1] If they do not dissolve completely, add more potassium hydroxide solution. Do not add more than another 25 mL, or your mixture will not fit in the separatory funnel!

[2] S. Budavari, M.J. O'Neil, A. Smith and P.E. Heckelman, eds. *The Merck Index, 11ᵗʰ Ed.* Rahway, NJ: Merck & Co., Inc. (1989). Based on several years of lab reports, it is not obvious to everyone that the specific rotation of (+)-MBA is +40.3°, while that of (−)-MBA is −40.3°.

For the report

Show all reactions involved in this experiment in your report. Report and discuss all observations. Broken down:

- Report three yields, as percentages:

  o MBA tartrate from MBA + tartaric acid

     ▪ In order to do this you must determine which (MBA or tartaric acid) is the limiting reagent, and know the formula weight of MBA tartrate.[1]

  o MBA from MBA tartrate

  o Final MBA as a percentage of initial MBA

- Report optical purity (e.e.) of your final product, as well as the mole fractions of (+) and (-) MBA enantiomers.

- Briefly discuss the chemical reactions you used to get from racemic MBA to optically active MBA.

Finding optical purity and mole fraction of enantiomers

Optical purity

Optical purity (e.e., "enantiomeric excess") is defined as the percentage relationship between the observed optical rotation and that which would be expected if the compound were all one enantiomer. It may be found by the following equation:

$$ e.e. = \frac{observed\ rotation}{[\alpha]_D \times length(dm) \times concentration(g\,/\,mL)} \times 100\% $$

where $[\alpha]_D$ is the specific rotation of the substance.

Mole fraction of enantiomers

The mole fraction of enantiomers may be found from the optical purity. Remember that each molecule of the (+) enantiomer cancels one molecule of the (-) enantiomer; any rotation left over stems from the presence of one enantiomer *in excess* of the other. The lower mole fraction is found by

$$ \frac{100\% - e.e.}{2}; $$

the higher mole fraction is found by adding the e.e. to the lower mole fraction. (Think about it, and you will see that this makes sense.)

For example, suppose you took 10g of a compound with a specific rotation of 100° and diluted it to 15 mL. The concentration would be 0.67 g/mL. Now, using a 0.95-dm cell, you observe a rotation of +8.4°. The optical purity would be found by

---

[1] The molecular weights of MBA (also known as 1-phenylethanamine) and tartaric acid are available in standard references. The formula weight of MBA tartrate may be found by inspecting the reaction scheme on page 1 and applying the numbers you just looked up.

$$e.e. = \frac{8.4}{100 \times 0.95 \times 0.67} \times 100\% = 13\%$$

The (+) enantiomer is present in excess over the (-) enantiomer (since we observed a (+) rotation). Therefore the mole fraction of the (-) enantiomer is (100-13)/2 or 43.5%; the mole fraction of the (+) enantiomer is 43.5+13=56.5%.

Schedule for "Resolution of a Racemic Mixture"

Week 1

1. Perform the initial steps in "Resolution of a Racemic Mixture."

You will leave your product to crystallize until next week. Be sure to label your flask properly!

2. *Orientation to the polarimeter.* You will learn to take readings, and determine the concentration (in grams/mL) of an unknown solution of an enantiomerically pure chiral compound.

Read the section on polarimetry in Karty Chapter 5 **before coming to lab**. The concentration of an optically-pure substance may be determined by measuring its *optical rotation*, or the extent to which it rotates plane-polarized light. To determine this, you must look up the compound's *specific rotation* $[\alpha]_D$ in a standard reference,[1] and use the equation

$$[\alpha] = [\alpha]_D \times d \times c$$

where $[\alpha]$ stands for the observed (measured) optical rotation of the solution, d stands for the path length in **deci**meters (not meters or centimeters!), and c is the concentration in **grams per milliliter**.

3. *Making molecules,* an exercise using your molecular model kit.

Week 2

Perform the steps in "Procedure, Second Week."

Thin-Layer Chromatography: Analysis of Analgesics

Chromatography is the most useful and widely applicable method available for separating chemical substances. There are many kinds of chromatography, including paper chromatography, column chromatography, flash chromatography, reverse-phase chromatography, ion exchange chromatography, gel permeation chromatography, high pressure liquid chromatography (HPLC), gas chromatography (GC), and so on.

All varieties of chromatography operate on the same principle: the *partition* of material between a *stationary phase* and a *mobile phase*. A sample (such as a mixture of unknown compounds) is *adsorbed* on the stationary phase (sometimes contained in a *column*), and the mobile phase is allowed or forced to flow past the sample. Each compound spends part of its time in the mobile phase, and part on the stationary phase. The amount of time spent dissolved in the mobile phase or adsorbed on the stationary phase is characteristic of each compound, and determines how fast that compound moves through the system.

---

[1] Specific rotations are always reported in the reference literature as positive numbers. However, levorotary compounds (negative rotation) will have a negative specific rotation, of course!

Read more about chromatography in general, and about TLC in particular, in Zubrick, Chapters 26 and 27, or one of the available organic laboratory manuals in the lobby of Shoker Science Center.

In this experiment, we will perform *thin layer chromatography* (TLC). TLC is a powerful and simple tool for identifying, determining the purity, and sometimes for purifying a sample of an organic compound (*preparative* TLC). **While Zubrick explains how to make your own TLC plates, we will be using commercially-prepared plates. Therefore you will not have to prepare TLC plates.**

In TLC, a thin layer of an adsorbant, such as silica gel or alumina, is bound to a solid support, such as a glass plate or a plastic sheet, forming the stationary phase. Silica gel is hydrated silicon dioxide ($SiO_2 \cdot nH_2O$, in which every surface oxygen atom is an OH group), and compounds that are capable of hydrogen bonding[1] will bind to it tightly. The less polar a compound is,[2] the less strongly it will bond to silica gel and the faster it will move up the plate.

In a typical procedure, a spot of the sample is applied near the bottom edge of the TLC plate. The plate is then placed vertically with the bottom edge immersed in a solvent.

NOTE: The spot should be completely out of the solvent pool!

The solvent flows up the plate by capillary action, serving as the mobile phase. The organic compounds in the solvent will partition between the mobile and stationary phases to a degree characteristic of each, depending on

* The polarity of the compound (more precisely, its tendency to form hydrogen bonds) and

* The polarity of the solvent (more precisely, its ability to act as a donor or acceptor of hydrogen bonds).

The more tightly the compound binds to the silica gel, the more polar the solvent must be to move it along the plate.

The plates we will use contain a fluorescent indicator, so that your spots will glow under ultraviolet light. After developing the plate, you should mark your spots with a pencil so that the plate can be read without using a UV lamp.

You must hand in the pre-lab to be admitted to the laboratory.

Experimental strategy

Analgesics are drugs that relieve pain. Some analgesics are narcotics, but not all: the first non-narcotic analgesic, aspirin, was extracted from willow bark.

Different over-the-counter analgesic drugs often contain the same active ingredients, and in some cases also contain caffeine. The structures of the common ingredients are shown in the table below.

You will run at least four plates, one in each of the solvents provided. Each plate will be prepared with five spots:

---

[1] Look up hydrogen bonding in Karty.

[2] Strictly speaking, binding to silica gel measures the ability to form hydrogen bonds. Even a relatively non-polar molecule will bind strongly to silica gel if it has several sites that can hydrogen-bond.

- One spot of each of the four standard solutions: aspirin, acetaminophen, ibuprofen, and caffeine.
- One spot that is a solution of your unknown, over-the-counter analgesic.

In addition to the four ingredients we are testing for, there may be other ingredients in commercial analgesics. These will appear as unidentified spots on your TLC plate.

Minimum safety standards for this experiment

1. Hot glass looks the same as cold glass. Be sure a piece of glass is cool enough to touch before trying to pick it up!
2. Ethyl acetate containing 5% acetic acid is flammable and smelly, but not particularly toxic. It must be used only in the hood.
3. Dichloromethane is a low-boiling, non-flammable solvent that can cause chemical intoxication in large quantities. It can also carry anything dissolved in it through your skin. Use dichloromethane only in the hood, and be careful not to allow any solutions to touch your skin.
4. Reagent alcohol is non-potable! It is moderately flammable. It can also carry anything dissolved in it through your skin. Be careful not to allow any solutions to touch your skin.
5. Acetylsalicylic acid, acetaminophen, ibuprofen and caffeine are toxic in large quantities, but present no concern in the quantities we are using.

Disposal

Acetaminophen	May be disposed of in the trash or flushed down the sink.
Acetone	Flush down the sink. Large spills may be cleaned with a paper towel and disposed of in the trash. Small spills should be allowed to evaporate.
Acetylsalicylic acid	May be disposed of in the trash or flushed down the sink.
Caffeine	May be disposed of in the trash or flushed down the sink.
Diethyl ether	Small quantities (less than 20 mL) should be poured out onto a paper towel and allowed to evaporate. The paper towel may then be disposed of in the trash.
Hexane	Small quantities (less than 20 mL) should be poured out onto a paper towel and allowed to evaporate. The paper towel may then be disposed of in the trash.
Ibuprofen	May be disposed of in the trash or flushed down the sink.
Methanol	Flush down the sink. Large spills may be cleaned with a paper towel and disposed of in the trash. Small spills should be allowed to evaporate.
Over-the-counter analgesics	May be disposed of in the trash or flushed down the sink.
TLC plates and silica gel powder	Dispose of in the trash.

Procedure

1. Mark four TLC plates **with pencil**. Lightly draw a line about a centimeter (½ inch) from the bottom of each plate, and mark this starting line with five evenly-spaced ticks to guide you in spotting the plate. Label each tick appropriately, so that you know which spot is which. Be sure to record the meaning of any abbreviations (such as "A" for "aspirin") in your lab notebook.

2. Spot the plates, *using a different capillary tube for each solution*, **at the starting line** (not above or below the starting line!) Try to place each spot in the center of the appropriate cross, made by the intersection of one of your tick marks and your starting line. **Each plate** should be spotted with your particular unknown solution, and with solutions of **all four** known compounds. Examine the spotted plates under the UV lamp to make sure that you have enough material to see.

3. Prepare the developing chamber for each of your plates, according to the procedure outlined in Zubrick. Each developing chamber should have a different solvent: acetone, diethyl ether, hexane or methanol.

4. Develop each of your plates, being sure to **mark the top of the solvent with pencil** immediately after removing the plate from its developing chamber. Allow your plates to dry.

5. Examining each plate under the UV lamp, outline (or mark the "front" of) each of your spots with pencil.

    a. **NOTE**: Like all "instrument readouts," your TLC plates, once developed and marked, should be fastened into your laboratory notebook. Do not use staples! Alternatively, tape or paste a photocopy of your TLC plates in your notebook.

    b. Turn in a **PHOTOCOPY** of your TLC plates with your report.

For the report

Report the R_f values for each of the known compounds in each solvent. **Use a method of presentation other than embedding numbers in a paragraph of text** (e.g. a well-organized table). Report the ingredients you found in your unknown, and identify the over-the-counter analgesic that was your unknown, based on its ingredients.

Try to correlate the R_f value for each of the four ingredients with its structure. Remember that centers of negative charge can participate in hydrogen bonding with silica! Which known compound had the strongest interaction with silica?

Which solvent resolved all the analgesics? Explain why, on the basis of the solvent's structure and polarity. Sort the solvents in order of polarity, and justify your choice on the basis of their structure and your observations in this experiment.

Pre-laboratory questions

1. Explain why you need to allow a TLC plate to dry completely after spotting, before either double-spotting or developing the plate.

2. Explain why you mark the TLC plate in pencil rather than in pen. (Zubrick should have something about this.)

3. You will be using acetone, diethyl ether, hexane and methanol as developing solvents. On the basis of structure, sort them in the expected order from most polar to least polar. Explain your reasoning.

4. Which type of fire extinguisher would you need to put out the most likely type of fire in this experiment? (Choose from water, CO_2, ABC dry chemical, or Type D. More than one correct answer may exist.) Justify your answer.

5. Are there any extraordinary hazards present in this experiment? Defend your answer. Consider not only the chemicals used, but also their amounts and the conditions.

Common analgesics and their active ingredients

Active ingredient	Structure	Which brand names contain these ingredients?
aspirin acetylsalicylic acid		Bayer Bufferin Anacin Excedrin
acetaminophen *N-para*-hydroxyphenyl-acetamide		Tylenol Datril Excedrin Panadol
ibuprofen 2-(4-isobutylphenyl)propanoic acid		Advil Motrin Brufen Nuprin
caffeine *N,N,N*-trimethylxanthine		No-Doz Anacin Excedrin

Introduction to two-dimensional NMR spectroscopy

The most important new tool introduced to organic chemistry in the past couple of decades is two-dimensional (2D) NMR spectroscopy, although we ourselves will not be able to make as much use of it as we would with a higher-resolution NMR spectrometer.[1]

2D NMR requires programming rather sophisticated experiments into the instrument. Our spectrometer is programmed for two kinds of 2D NMR: COSY (COrrelated SpectroscopY) and HETCOR (HETero-nuclear CORrelation).

- COSY is a ^{1}H experiment. It correlates two 1D ^{1}H NMR spectra, showing which peaks are coupled to which, as explained below. A COSY spectrum can be obtained in about 15 minutes, at one scan per slice; or 45 minutes, at four scans per slice.[2]

- HETCOR is a ^{13}C experiment, and thus requires much more sample than COSY. HETCOR correlates a ^{13}C spectrum, on the horizontal axis, with an ^{1}H spectrum, on the vertical axis, and is a way of determining which ^{13}C peaks correspond to which ^{1}H peaks. HETCOR is time-consuming and troublesome to obtain, and we will use it only as a last resort.

Using 2-dimensional NMR to solve structures

In 2D NMR spectroscopy, signal correlation can be seen at a glance. Two 1D spectra are plotted against each other, and which peaks are coupled to which can be read off the 2D plot. For example, consider ethylbenzene.

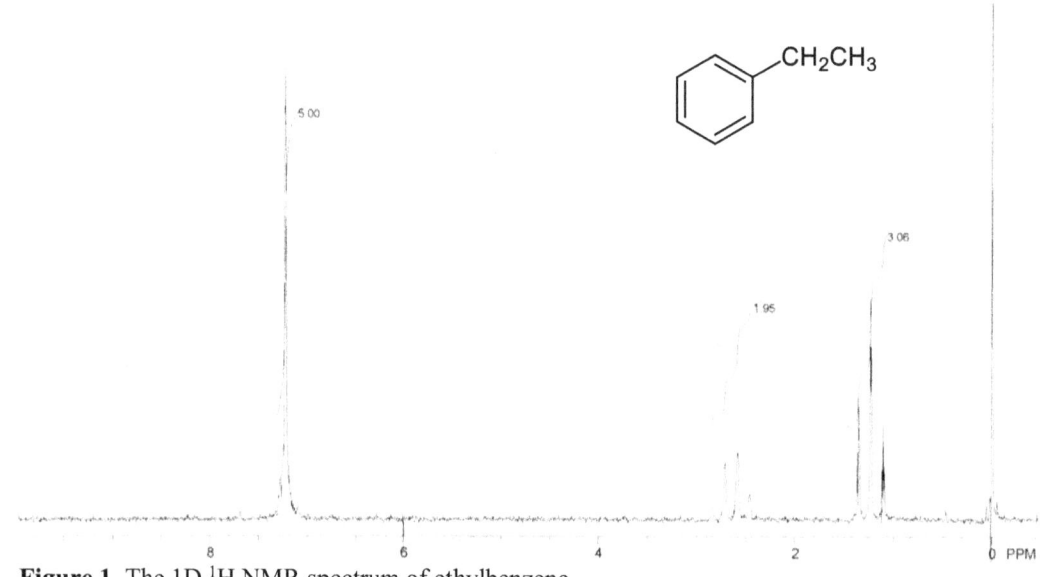

Figure 1. The 1D ^{1}H NMR spectrum of ethylbenzene.

In Figure 1, we see the 1-dimensional ^{1}H NMR spectrum of ethylbenzene. The five phenyl protons appear at 7.2 ppm, while the ethyl group is represented by the 2H quadruplet at 2.6 ppm and the 3H triplet at 1.2 ppm. The correlation of the CH_2 and CH_3 are obvious to those who know how to read splitting patterns,

---

[1] At 60 MHz, ^{1}H NMR peaks often overlap too much for good 2-dimensional resolution.

[2] The more time-consuming 4-scan COSY is less prone to detecting false peaks, which are especially troublesome in the spectra of esters and other oxygen-containing molecules.

70

but there is another, easier way to establish that the hydrogens at 2.6 and those at 1.2 are vicinal to each other: the 2-dimensional COSY spectrum.

In a COSY spectrum, two identical ¹H NMR spectra are correlated. Spots on the diagonal from upper right to lower left represent self-correlation, the intersection points of identical peaks; but off-diagonal spots represent correlations between different hydrogens in the spectrum. In Figure 2, observe the couplings between the peaks at 1.2 and 2.7 ppm (circled). It is immediately apparent that the hydrogens represented by those peaks are vicinal to each other because their coupling produces an off-diagonal peak; unlike the 1D spectrum (Figure 1), no interpretation of splitting patterns is required. Notice that long-range coupling, such as that between the phenyl hydrogens and the ethyl group, is not seen in this spectrum; in general it is not found in 2D NMR. (The coupling peaks to 5.7 ppm are spurious.)

A slightly more difficult example, at least at low resolution (60 MHz), is 2-butanone.

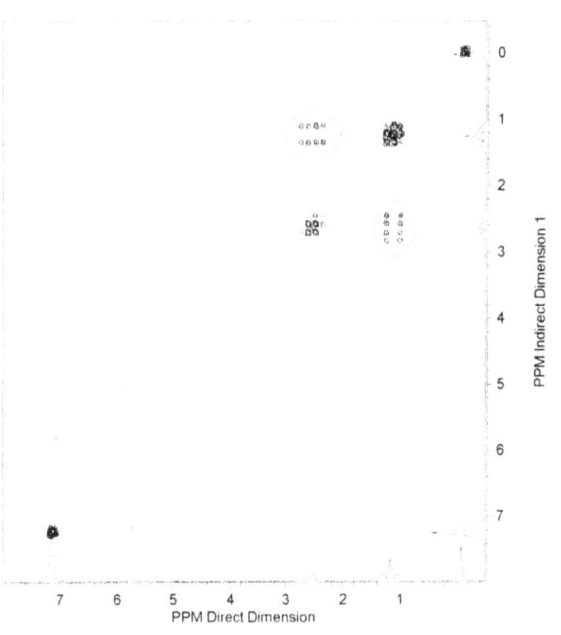

Figure 2. The COSY spectrum of ethylbenzene. Circled peaks are correlations between the peaks at 2.7 and 1.1 ppm.

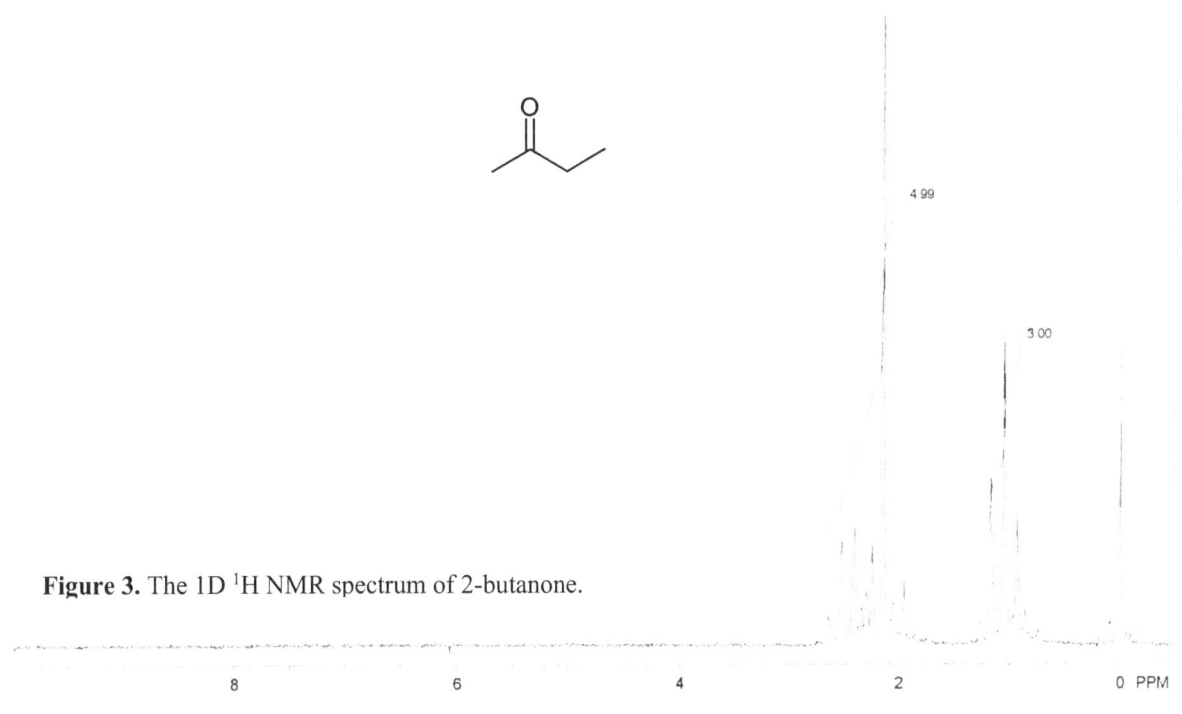

Figure 3. The 1D ¹H NMR spectrum of 2-butanone.

In Figure 3, it's easy to see the triplet at 1.0 ppm (although it is obscured by sideband peaks); the integration is 3, which means that it corresponds to three hydrogens coupled to two vicinal hydrogens. But the peaks from 2.0-2.7 ppm are more difficult. Even discounting the sidebands, the large singlet from

71

the COCH₃ hydrogens overlaps the quadruplet from the $COCH_3$ hydrogens overlaps the quadruplet from the $COCH_2$, which are coupled to the methyl triplet. While the integration is a tremendous aid to interpretation of this spectrum, it is possible to miss the triplet-quadruplet pattern due to the overlap of the other methyl group.

But in the COSY spectrum (Figure 4), matters become easier. Notice that the large peak at 2.2 ppm (a) is not coupled to any other hydrogens, but there is a peak (b) from 2.2-2.6 that is strongly coupled (c) to the peak (d) at 1.1 ppm.[1] The 2D spectrum allows us to disentangle the coupling patterns and show that we do, in fact have independent CH_3— and CH_3CH_2— groups.[2]

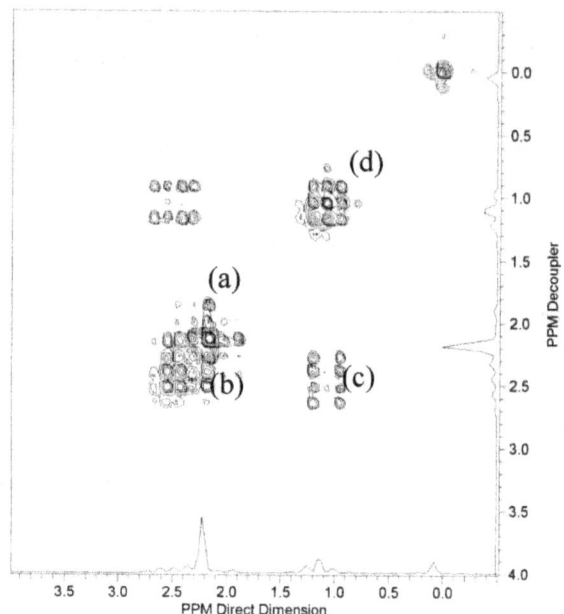

Figure 4. The COSY spectrum of 2-butanone. The large peak at 2.2 ppm (a) is not coupled, but the peaks from 2-2.6 (b) and 1-1.3 (d) are coupled at (c).

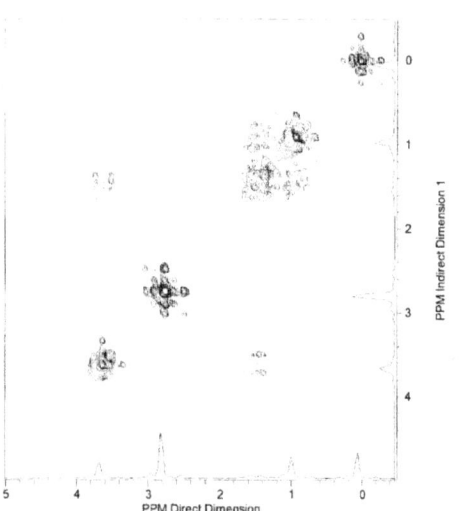 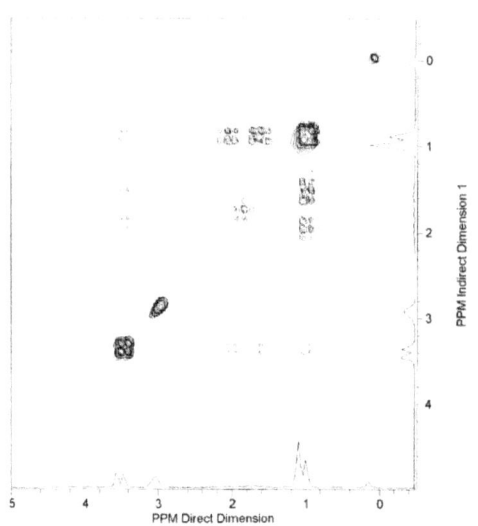

Figure 5. COSY spectra of 1-butanol (left) and 2-methyl-1-propanol (right). Find the correlation patterns. Don't be confused by the OH peak; that hydrogen exchanges too rapidly for correlation, and comes around 3 ppm.

Now compare the COSY spectra for 1-butanol and 2-methyl-1-propanol (isobutyl alcohol), in Figure 5.

---

[1] Notice how the triplet/quadruplet pattern is preserved in the cross peaks.

[2] Note also that we must use both 1D and 2D spectra to obtain all available information: COSY doesn't do integration!

The limitations of 60 MHz resolution are immediately apparent. There is no separation between the C2 and C3 methylene groups in the spectrum at left, and so it is difficult to see the different coupling between *n*-butyl alcohol (Figure 5, left) and isobutyl alcohol (Figure 5, right). However, we are able to see long-range coupling between the C1 methylene and the physically close terminal methyl groups (C3) in the right-hand spectrum, while this coupling is not present in the left-hand spectrum. With the presence of two doublets in the 1D spectrum of 2-methyl-1-propanol (easily visible in the right-hand spectrum), this allows us to distinguish it from 1-butanol.

As a further exercise, consider the two COSY spectra shown in Figure 6. One is 2-phenylbutanoic acid, the other 3-phenylbutanoic acid. Can you tell which is which?

Figure 6. Which COSY spectrum is 2-phenylbutanoic acid, and which is 3-phenylbutanoic acid? Answer on page 75. Aryl and acid hydrogens are not included in the COSY spectra.

Virtual Spectroscopy[1]

Student: _____

You will use the Integrated Spectral Database System for Organic Compounds (SDBS)[2] for this assignment. Compound numbers refer to the internal reference number for each compound on SDBS.

You have been assigned the primary compound number _____ and five companion compounds

numbered _____, _____, _____, _____ and _____.

1. Look up the primary compound on SDBS.

2. Draw the structure (given on SDBS).

3. Print the four spectra: MS, IR, ^{1}H NMR and ^{13}C NMR.

4. For each companion compound:

   a) Look up the compound on SDBS.

   b) Identify structural features that would be expected to spectroscopically distinguish the companion compound from your primary compound. For example, the companion may have a carbonyl group where the primary does not.

   c) View the spectra of the companion compound. Choose one distinguishing spectroscopic feature, and write a brief explanation of how that feature is inconsistent with the structure of your primary compound. You are expected to choose a feature that would convincingly differentiate the companion from the primary compound. For this reason, it is most likely that you will need to use several different spectroscopies for your different compounds. See the examples below.

   Example. The IR of compound #2428 1,4-benzodioxane shows no peak in the carbonyl region 1680-1740 cm^{-1}. My compound #1725 benzyl formate has a C=O group which absorbs strongly at 1725 cm^{-1}.

   Example. The ^{1}H NMR of compound #1448 *para*-toluic acid has a methyl group at $\delta = 2.375$. My compound does not have a methyl group and shows no peaks in that region.

   Example. The MS of compound #725 methyl benzoate shows the base peak at 105, which is 31 mass units from the molecular ion peak at 136. This corresponds to loss of a methoxy group. My compound does not have any methoxy groups and shows only a small peak at 105.[3]

---

[1] Kandel, M.; Tonge, P.J. *J. Chem. Ed.* **2001**, *78*, 1208-1209.

[2] http://riodb01.ibase.aist.go.jp/sdbs/cgi-bin/cre_index.cgi, maintained by the Japanese National Institute of Materials and Chemical Research.

[3] Lest you think this is an easy way out, be advised that all your compounds are constitutional isomers!

Answer to Figure 6 on p. 73

The left-hand spectrum in Figure 6 is 3-phenylbutanoic acid; the right-hand spectrum is 2-phenylbutanoic acid. How do the correlation patterns allow you to choose between the two structures?

Qualitative Analysis: Identification of Unknown Organic Compounds

Scenario

You work for a chemical testing laboratory. A client has sent in several samples for identification, from reagent bottles that have lost their labels. All that is known about these samples is that they are organic chemicals; they must be identified before a waste hauler will take them away. You must identify the unknown organic substances from their melting or boiling points and mass, IR, and NMR spectra.[1]

The experiment

You will be given two organic samples identified only by number. The mass spectrum of each sample will be provided to you. You will obtain the boiling or melting point, the IR spectrum and any NMR needed[2] for full structural identification of each sample. You will identify each sample using the data you have obtained.

Your report will fully discuss the identification of each of your unknowns.

Week 1 of the experiment

You will be given your unknowns this week, along with their mass spectral and CHO elemental analysis data. Obtain the IR spectra of your unknowns, and submit a brief, informal report outlining the functional groups found. This report (with spectra attached) is due the following week.

Last week of the experiment (indicated on the lab schedule)

During the scheduled laboratory periods during this week you will have the opportunity to obtain any spectra you still need. The instructor will be available during the day to assist you in obtaining good spectra.

Note that a failure to obtain good hydrogen NMR spectra – that is, spectra with well-resolved peaks rather than amorphous blobs – will cost you a significant number of points. Don't be lazy; ask for help!

---

[1] This really happened to Dr. Berger in graduate school!

[2] This may include ^{1}H, ^{13}C, DEPT, COSY, or HETCOR spectra.

For the report...

Discuss your spectra and other data **in detail** as they relate to your structural assignment. It is legitimate to use standard reference sources such as the Aldrich spectral libraries or SDBS, but you should credit them in your report if you do. It is NOT sufficient to simply state that your spectrum matches an authentic spectrum! You must discuss the structural elements indicated by your spectra!

Minimum Safety Standards for this experiment

Since the compounds you will be dealing with are unknowns, it is not possible for you to be certain of their toxicity. Therefore, you must treat ALL of the unknowns with special precautions. This includes working in the hood when possible and avoiding contact with the liquids and their vapors.

The hoods MUST be on when you are obtaining micro melting points or preparing samples for spectroscopy. Safety glasses MUST BE WORN during sample preparation and when obtaining micro melting or boiling points.

Disposal and Procedure

You are permitted to prepare samples and perform spectroscopy, or boiling- and melting-point determinations, any time there is a science professor in the building. A professor does not need to be present in the laboratory.

Micro boiling points should be obtained as detailed in Zubrick. Clean the boiling point tubes with acetone, NOT soap-and-water.

When obtaining **IR spectra**, use the round windows (with the 25-μm spacer) for liquid samples and the thick "mull plates" for solid samples. Solid samples must be ground fine and mixed with a small amount of mineral oil (a "Nujol mull"). IR cell parts should be cleaned with dry isopropyl alcohol.

To obtain **NMR spectra**, you will use NMR tubes. NMR samples must be properly labeled, including the date prepared. Samples without proper labels will be discarded without warning. You are allowed to store prepared samples in the NMR tube, in the rack in the instrument room. However, no more than two NMR samples per student are allowed. We do not have infinite supplies of NMR tubes.

Mix your NMR samples with chloroform-*d*. **If your unknown is insoluble in chloroform, you will be told what solvent to use.** Solid samples should be mixed in a vial to ensure solubility before putting in the NMR tube! Floating solids will make your spectra unusable! If you are unable to dissolve all your solids, filter your sample into your NMR tube, through a small amount of cotton waste in a Pasteur pipet. See Zubrick, pp. 69-70ff.

Samples for ^{13}C NMR must have a concentration of at least 50% by weight[1] in chloroform-*d*. A good ^{1}H spectrum may be obtained with any concentration over 5% by weight, or even less if you average several (no more than 20!) scans. For best results, NMR samples should have a total volume of ca. 5 mL; see the NMR sample height guide, posted by the hood.[2]

---

[1] Liquid samples should contain about 0.3 mL of the unknown and 0.2-0.3 mL of chloroform-*d*. For solid samples, weigh about 0.7 mmol of the solid and mix with 0.7 mL of chloroform-*d*. Samples should be prepared in a small vial and filtered into the NMR tube through cotton waste.

[2] If the NMR sample is not tall enough, you will NOT get a good spectrum!

Chloroform solutions (i.e. NMR samples) must be placed in the labeled waste bottle provided. This will remain true for the entire term. The waste bottle for NMR samples may be kept in the hood or on the benchtop as long as it is tightly closed.

Clean your NMR tubes with isopropyl alcohol or acetone and store them **upside down** (to prevent dust contamination) in the beaker provided.

Rinsings may be flushed down the sink. Check the sink afterward for odors!

Identification of an unknown C₇ ester

This experiment will use all the spectroscopic knowledge you have.

You will synthesize an unknown ester $C_7H_{14}O_2$ from an acid and an alcohol that will be supplied to you already mixed. The acids and alcohols will split the seven carbons evenly, so that 3-carbon acids will be paired with 4-carbon alcohols, and 4-carbon acids with 3-carbon alcohols.

You will purify the ester and identify it by NMR spectroscopy. While the experiment is designed to give you enough product for ^{13}C NMR, a combination of 1D and 2D ^1H NMR should be sufficient (though DEPT will probably be helpful, and doesn't require as much material as a standard ^{13}C spectrum). You may use simulated or recorded one-dimensional spectra[1] to help solve your structure.

HINTS: Look for patterns in your spectra that allow you to zero in on a particular set of structures. Pay particular attention to the hydrogens on the carbon bonded to oxygen (that is, the O-CH₂ or O-CH), if any; where do such hydrogens typically appear? What about the hydrogens on the carbon next to the carbonyl group? Where do such hydrogens typically appear? What does the splitting tell you? What does the integration in the 1D hydrogen spectrum tell you?[2]

You will have a week to solve the structure of your ester. Ask for help if you need it!

Techniques used from Zubrick, 8th Ed: reflux (Chapter 22), extraction and washing (Chapter 15), IR spectroscopy (Chapter 32), NMR spectroscopy (Chapter 33)

Safety and disposal guidelines

Nothing you will use in this experiment is more than an irritant, except for concentrated sulfuric acid which is quite corrosive. However, the organic acids and esters are smelly;[3] be thorough when disposing of them and ensure that, for example, pipets are cleaned before discarding so that no odor remains.

Procedure

1. You will be given a mixture of an unknown alcohol (approx. 50 mmol) and an unknown acid (approx. 60 mmol) in a 25-mL round-bottomed flask. Add about ½ mL of concentrated sulfuric acid. Add boiling chips. Clamp the flask so that it is half-submerged in a beaker of water, and attach a condenser; cap the condenser with an empty thermometer adapter to minimize the ingress of water vapor. Run water through the condenser and heat your mixture in a hot/boiling water bath, for one hour.

---

[1] Look at the SDBS and at nmrdb.com for spectra.

[2] Be careful! Integration may not be as useful as splitting patterns.

[3] "Smelly" can be rather pleasant with the esters we are using, but the odor will be strong.

2. At this point you will have two layers. What are they? (What did you put in? What is your product? Do you expect to have anything left over?) Separate the layers and discard the one that is not your product.

3. Wash your product three times with 10-mL portions of 5% sodium carbonate[1] and once with 10 mL of saturated sodium chloride. Dry the organic layer over calcium chloride in a large test tube or small flask, whichever is appropriate for the volume you have.

4. Obtain the IR spectrum of your product. If you have an excessive amount of alcohol or acid remaining, repeat the Na_2CO_3 and NaCl washings. It is important that your ester be reasonably pure, or you will not get good NMR results!

Identifying your product

Use NMR spectroscopy – taking advantage of COSY and ^{13}C methods – to identify your product. The instructor will be available to help you obtain good spectra. Ask for help with analysis if you need it.

For the report

The report must discuss, in detail, how you identified your ester and show the structures of your alcohol and acid starting materials. Draw the mechanism for the reaction, showing your specific acid-alcohol combination.

Do not bother to calculate a yield, as you are not being told the precise masses that you started with.

Pre-laboratory assignment

1. What are the structures of the esters with formula $C_7H_{14}O_2$?[2] Think about how you would tell them apart by NMR.

2. Are there any extraordinary hazards present in this experiment? Defend your answer. Consider not only the chemicals used, but also their amounts and the conditions.

Identification of an unknown C₁₄ ester[3]

This experiment will use all the spectroscopic knowledge you have, along with a number of tools used in the "real world" such as spectroscopic simulation.

You will synthesize an unknown ester from an acid $C_{10}H_{12}O_2$ and an alcohol $C_4H_{10}O$.[4] What is the formula of the resulting ester?

---

[1] What is the purpose of the sodium carbonate? HINT: what is the stoichiometry of the reactants you were given?

[2] That is, esters of 3-carbon acids $C_3H_6O_2$ with 4-carbon alcohols $C_4H_{10}O$, and esters of 4-carbon acids $C_4H_8O_2$ with 3-carbon alcohols C_3H_8O.

[3] This experiment is based on S.E. Branz, R.B. Miele, R.K. Okuda and D.A. Straus, *J. Chem. Educ.* **1995**, *72*, 659-661.

[4] The acids and alcohols used are purchased from Aldrich. This should allow you to narrow the range of possibilities.

You will purify the ester and identify it by NMR spectroscopy. A combination of 1D and 2D ^{1}H NMR should be sufficient, though you should have enough for ^{13}C NMR and DEPT. You may use simulated one-dimensional spectra to help solve your structure; however, you must ask for the spectrum of a particular ester, and you are only allowed a total of three simulated spectra.

HINTS: Look for patterns in your spectra that allow you to zero in on a particular set of structures. Pay particular attention to the hydrogens on the carbon bonded to oxygen (that is, the O-CH$_2$ or O-CH), if any; where do such hydrogens typically appear? What does the integration in the 1D spectrum tell you?[1]

Ask for help if you need it!

Techniques used from Zubrick, 8th Ed: reflux (Chapter 22), extraction and washing (Chapter 15), IR spectroscopy (Chapter 32), NMR spectroscopy (Chapter 33)

Safety and disposal guidelines

Nothing you will use in this experiment is more than an irritant, except for concentrated sulfuric acid which is quite corrosive. However, the acids and especially the esters are smelly; be thorough when disposing of them and ensure that, for example, pipets are cleaned before discarding so that no odor remains.

Procedure

1. You will be given a mixture of an alcohol (1.0 g of $C_4H_{10}O$) and an acid (2.5 g of $C_{10}H_{12}O_2$) in a 25-mL round-bottomed flask. Add about ¼ mL of concentrated sulfuric acid, about 1 mL of acetonitrile, and one or two boiling chips. Attach a condenser; cap the condenser with a thermometer adapter, to allow pressure equalization while minimizing exposure to atmospheric moisture. Heat the mixture in a boiling water bath for 60 minutes.

2. Add 10 mL of ether, mix, and separate any water layer that may have formed. Wash the organic layer twice with 5-mL portions of 5% sodium carbonate and once with 5 mL of saturated sodium chloride. Dry the organic layer and remove the ether using a rotary evaporator.

3. Obtain the IR spectrum of your product. If you have an excessive amount of alcohol or acid remaining, dissolve in ether and repeat the Na$_2$CO$_3$ and NaCl washings. It is essential that your ester be reasonably pure, or you will not get good NMR results![2]

4. Use NMR spectroscopy to identify your product; COSY and DEPT will probably be helpful. The instructor will be available to help you obtain good spectra. Simulated 1D ^{1}H NMR spectra will be supplied on request; you must ask for the spectrum of a specific $C_{14}H_{20}O_2$ ester. You are limited to three simulated spectra.

---

[1] Be careful! Past results have shown that students may get a mixture of the ester and the starting acid, so that integration may not be as useful as splitting patterns. If an acid peak appears in the one-dimensional ^{1}H NMR, its integration may tell you what proportion of your product is ester and what proportion is acid, which should help in peak assignments.

[2] You may get some unreacted acid in your product; this will throw your integration off!

For the report

The report must discuss, in detail, how you identified your ester and show the structures of your alcohol and acid starting materials. Draw the mechanism for the reaction, showing your specific acid-alcohol combination.

Pre-laboratory assignment

1. Draw the structures of the possible alcohols ($C_4H_{10}O$). How would you distinguish them by ^{1}H NMR?

2. The acids used for this experiment are shown below. How would you distinguish the acids by ^{1}H NMR (consider both one-dimensional and COSY)?

3. Are there any extraordinary hazards present in this experiment? Defend your answer. Consider not only the chemicals used, but also their amounts and the conditions.

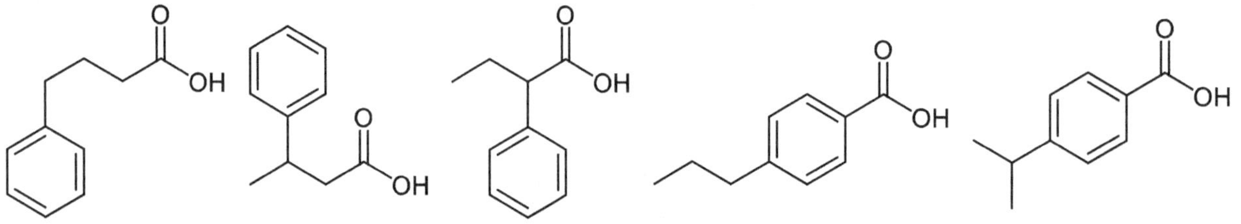

Reaction Studies

The following experiments are intended as introductions to the methods used to elucidate how organic chemical reactions happen. Mechanistic organic chemists use these methods, and others like them, to determine the steps and intermediates in chemical reactions and the structural features of molecules that influence reactivity.

Not all of these experiments will be used in any given academic year.

Factors affecting the reactions of alkyl halides

The S_N1, S_N2, E1 and E2 reactions are discussed in detail in Chapters 9-10 of Karty. In this experiment we will examine the effects of leaving group, alkyl group structure, and solvent polarity on the reaction of alkyl halides.

We will use reactions with sodium iodide in acetone and with silver nitrate in ethanol. We will also examine the effect of solvent polarity (40%, 50% and 60% 2-propanol in water) on the reaction of a tertiary alkyl halide with sodium hydroxide.

Note: Groups of two to three students will each perform one of the three experiments. Data will be reported to the instructor using a provided spreadsheet template, and the instructor will provide a complete set of all data to students.

Review Karty Chapters 9-10 before lab, including the effect of nucleophile and solvent on the reactions.

Minimum safety standards for this experiment

1. Glass that is wet with alcohol solutions may feel slippery. Be careful handling it.

2. The alkyl halides used in this experiment are inflammable, and some are irritants. Treat them with respect and wash your hands after handling.

3. 2-propanol is inflammable and an irritant; the solutions we will use are similar in composition to commercial rubbing alcohol. Treat them with respect.

4. The sodium hydroxide solutions you will handle will be relatively dilute. Wash your hands carefully and often.

5. Sodium iodide is an irritant. The amounts we will use present essentially no hazard.

6. Silver nitrate is mildly toxic, and will stain your fingers gray. Use care in handling it.

Disposal

All mixtures from this experiment may be flushed down the sink.

Procedure

Organic halides that may be used in this experiment:

1-chlorobutane	2-chloro-2-methylpropane	1-chloro-2-butene (crotyl chloride)
1-bromobutane	2-bromo-2-methylpropane	4-methylbenzyl chloride
2-chlorobutane	bromocyclopentane	bromobenzene
2-bromobutane	bromocyclohexane	

1. **Reactions with sodium iodide in acetone.** Sodium iodide in acetone causes S_N2 reactions exclusively. Why is that?

    a. Preheat a water bath to 50°C (what is the boiling point of acetone?) If your test tubes seem wet, rinse them with acetone.

    b. Label one test tube for each alkyl halide you will be testing. Place four drops (~0.2 mL) of each alkyl halide in the appropriate test tube.

    c. Add 2 mL of 15% NaI in acetone to each test tube and shake to mix. Note the time of mixing. Observe the test tubes for 5-10 minutes and note the time necessary for a precipitate to form.

    d. Take the tubes that have not formed a precipitate and heat them for 5 minutes in the 50°C water bath. Cool them to room temperature and note whether a precipitate has formed or not.

    e. If there is time, perform a second set of these tests. Clean all tubes with water and soap, if needed, and rinse with acetone from the squirt bottle.

2. **Reactions with silver nitrate in reagent alcohol.**[1] Silver nitrate promotes $S_N1/E1$ solvolysis by ethanol. Why is that?

    a. Preheat a water bath to 80°C (what is the boiling point of ethanol?) If your test tubes seem wet, rinse them with reagent alcohol.

    b. Label one test tube for each alkyl halide you will be testing. Place four drops (~0.2 mL) of each alkyl halide in the appropriate test tube.

    c. Add 2 mL of 1% silver nitrate in reagent alcohol to each test tube and shake to mix. Note the time of mixing. Observe the test tubes for 5-10 minutes and note the time necessary for a precipitate to form. (What is the precipitate?)

    d. Take the tubes that have not formed a precipitate and heat them for 5 minutes in the 80°C water bath. Cool them to room temperature and note whether a precipitate has formed or not.

    e. If there is time, perform a second set of these tests. Clean all tubes with water and soap, if needed, and rinse with reagent alcohol from the squirt bottle.

3. **Reactions of a tertiary alkyl bromide with sodium hydroxide in 2-propanol/water solutions.** What sort of reaction(s) do you expect from this combination of reagents?

---

[1] "Reagent alcohol" is a non-potable mixture of methanol, ethanol and isopropyl alcohol with the same dielectric constant as absolute ethanol.

a. To prepare 50% 2-propanol (a 50% solution of 2-propanol in water), place 50 mL of 2-propanol into a 100-mL graduated cylinder, then add water until the total volume is 100 mL. Mix well. 40% and 60% 2-propanol are prepared in a similar way.

    In the tests below, if the solution does not become colorless within 15 minutes, stop the reaction, record the time as "greater than 15 minutes" and do not repeat the experiment.

b. Place 50 mL of 50% 2-propanol in an Erlenmeyer flask containing a magnetic stirring bar. Add 5 drops of phenolphthalein indicator and exactly 200 μL of 0.5 M NaOH to the flask. Mix well. What color is the solution?

c. Add 50 μL of 2-bromo-2-methylpropane (*tert*-butyl bromide) to the flask, with swirling, and measure the time required for the solution to become colorless.

d. Repeat b and c using 40% 2-propanol, and using 60% 2-propanol.

e. If there is time, perform a second set of these tests.

f. If there is time, perform the test in your fastest solvent mixture, using 2-chloro-2-methyl-propane (*tert*-butyl chloride) as the alkyl halide.

Pre-laboratory assignment

1. Explain what type(s) of reaction each of the three sets of reaction conditions promotes: S_N1, S_N2, E1 or E2. Explain why.

2. Predict which of the organic halides is likely to react under each set of conditions.

3. Are there any extraordinary hazards present in this experiment? Defend your answer. Consider not only the chemicals used, but also their amounts and the conditions.

For the report

No experimental section is needed. **You must use the full data of all students for your report.**

All results will be reported individually, and -- if appropriate, meaning you have more than two or three data points for a particular reaction under particular conditions – as averages with standard deviation.

Discuss the observed effects (including primary/secondary/tertiary and other organic group effects; leaving group effects; and solvent effects) in terms of the theory of nucleophilic substitution and elimination reactions.

Are your pre-lab predictions borne out by the results you obtained?

^{1}H NMR analysis of keto-enol tautomerism[1]

All carbonyl compounds with hydrogens *alpha* to the carbonyl group exist in equilibrium with their enol forms. Normally the equilibrium strongly favors the keto tautomer. However, β-dicarbonyl compounds, in which two carbonyl groups are separated by a methylene (CH_2) group, have the possibility of hydrogen bonding between the enol of one carbonyl group and the neighboring carbonyl oxygen:

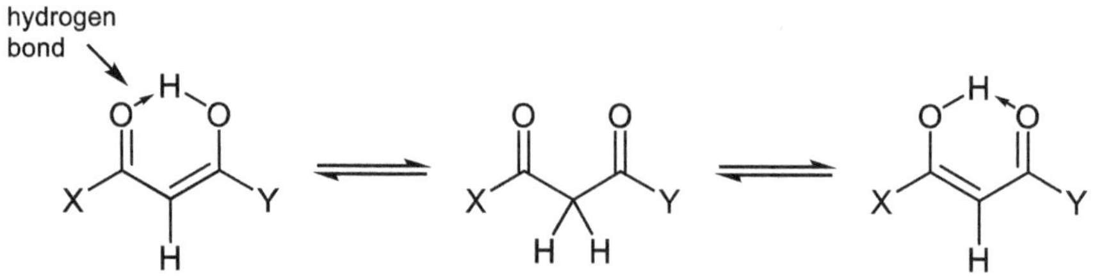

We can distinguish the keto and enol forms by ^{1}H NMR: while the saturated hydrogens (CH_2) between the carbonyl groups in the keto form will appear around 3.5-4.5 ppm, the vinylic hydrogen (=CH) of the enol form appears between about 5.0 and 6.5 ppm. Because the integration of a ^{1}H NMR peak indicates how many hydrogens correspond to that peak, we can determine the ratio of saturated to vinylic hydrogens by integrating the 1H NMR spectrum of each sample. This ratio will give us the equilibrium constant for conversion of the keto form to the enol form, according to the relationship[2]

$$K_{eq} = \frac{[enol]}{[keto]} = \frac{2 \times \text{enol integration}}{\text{keto integration}}$$

The equilibrium constant K_{eq} corresponds to the change in free energy ΔG^0 for conversion of the keto form into the enol form:

$$\Delta G^0 = -RT \ln K_{eq}$$

where R is the gas constant[3] and T is the temperature (assume 293 K, which is 20°C).

We will determine the enol-to-keto equilibrium constants[4] for several β-dicarbonyl compounds in chloroform-*d* at a concentration of approximately 0.3 moles per kilogram of chloroform-*d*. Samples will be prepared by the instructor.

Procedure

This experiment will be performed by lab section; each lab section will cooperate in obtaining the necessary spectra. However, **data analysis** must be performed *individually*. If different sections use different compounds, all data will be posted to the course website and all students are expected to use all of it.

1. Obtain at least five ^{1}H NMR spectra of each solution, including integration of the relevant peaks.

---

[1] Based on E.J. Drexler and K.W. Field, *Journal of Chemical Education* **1976**, *53*, 392-393. See also A. Grushow and T.J. Zielinski, *ibid* **2002**, *79*, 707-714.

[2] We need to multiply by 2 to allow for the fact that the keto form has two saturated hydrogens between the carbonyl groups, but the enol form has only one hydrogen in the corresponding position.

[3] Look it up. Use the one with units in kcal/mol for ease of comparison to your calculated results.

[4] It is STRONGLY suggested that you re-read the section on equilibrium constants from your General Chemistry text.

2. Analyze the NMR spectra to obtain the equilibrium constants and free energy differences for the interconversion of each tautomeric pair. You may assume that the probe temperature is 20°C unless otherwise instructed.

For the report

Report the equilibrium constant for each of the compounds we examined, and explain what it is that the equilibrium constants are measuring. What do they mean, and what do they tell us about preferred structures?

Explain any observed differences between the β-dicarbonyl compounds studied, in terms of the substituents of each. You may speculate, but be reasonable! For example, consider the effect of alkoxy groups on electrophilic aromatic substitution reactions; are they electron-donating or electron-withdrawing?

No experimental section is required in your report.

Pre-laboratory assignment

1. Look up the pK_a values for each of the following compounds: 3-oxobutanal, 2,4-pentanedione (acetylacetone), ethyl 3-oxobutanoate (ethyl acetoacetate) and diethyl propanedioate (diethyl malonate). Compare to the pK_a values for ethanol, 2-propanone and ethyl ethanoate.

2. Explain why β-dicarbonyl compounds are more acidic than monocarbonyl compounds.

Kinetics of the Esterification of Trifluoroacetic Acid[1]

We will use NMR to determine the pseudo-first-order rate constant for esterification of trifluoroacetic acid.

$$F_3CCOH + ROH \longrightarrow F_3CCOR + H_2O$$

Calculating the rate constant

The rate of esterification of any acid (seen as the rate of appearance of the ester or disappearance of the alcohol) will be dependent on the concentration of the acid and the concentration of the alcohol according to the differential equation

$$\frac{d[ester]}{dt} = -\frac{d[alcohol]}{dt} = k[acid][alcohol]$$

in which [x] is read "the concentration of x" and k is called the **rate constant**.[2] This equation tells us that both the acid and the alcohol are involved in the rate-limiting transition state for the reaction; we know this because the rate is dependent on both their concentrations. Since the sum of the exponents of the

---

[1] This experiment is based on D.E. Minter and J.C. Villarreal, *J. Chem. Ed.* **1985**, *62*, 911-912.

[2] The "rate constant" is not really a constant, as it varies not only with the specific reaction but with temperature.

concentrations is 2 (1 + 1), we call this a "second-order reaction." It is possible to solve a second-order rate law numerically, but it is much more difficult to solve it analytically.[1]

However, if the concentration of one reactant (in our case, the acid) is large compared to that of the alcohol, its concentration does not change significantly during the course of the reaction and we can define a "pseudo-first-order" reaction described by the differential equation

$$\frac{d[ester]}{dt} = -\frac{d[alcohol]}{dt} = k'[alcohol]$$

where the "pseudo-first-order rate constant" $k' = k[acid]$. This allows us to define the rate solely in terms of the concentration of the alcohol. Collecting terms and rearranging, we have

$$-\frac{d[alcohol]}{[alcohol]} = k'dt$$

which can be integrated thus:

$$-\int_{[alcohol]_0}^{[alcohol]_t} \frac{d[alcohol]}{[alcohol]} = \int_0^t k' \, dt$$

which reduces to

$$-\{\ln[alcohol]_t - \ln[alcohol]_0\} = k't \qquad \text{so that } [alcohol]_t = [alcohol]_0 e^{-k't}$$

The integrated rate equation is solved numerically (curve fitting) by the NUTS software, using ^{1}H NMR integration data over a series of scans. The software allows analysis of either the decay of the alcohol peak or the growth of the ester; when possible, we will do both. The output from the software is a number (T1 or T2) that is the **reciprocal** of the pseudo-first-order rate constant k' for the reaction.[2]

The goal of this week's experiment

All experiments will be run at a concentration of 1 molal, that is, 1 mmol of alcohol per gram of trifluoro-acetic acid.[3] Solutions will be prepared during the lab period; some data may be collected by the professor prior to lab, and supplied to you. Combined data will be posted to the course web page.

We will compare the rates of the esterification of trifluoroacetic acid by a series of alcohols. How do these alcohols differ in their substitution?

---

[1] "To solve analytically" means to be able to write an integrated equation for the concentrations of reactants as functions of time.

[2] We could also analyze raw data from the experiment by manually integrating each spectrum in the data set. However, because we can't assume that the integral response for the ester is the same as that for the alcohol, we can only use the alcohol data for this analysis.

If we were to use ester integration we would have to obtain an integral after the solution had stood for several hours—we call this "concentration of ester at infinite time" or $[ester]_\infty$—and the integration response would probably not be equivalent to that obtained during the experiment.

[3] Are these conditions acidic? How strong an acid is trifluoroacetic acid?

Data will be obtained under the supervision of the professor, and will be analyzed by you. Account for any differences in the observed rate constants. Are the differences significant? How can you tell?

For the report

Discuss the mechanism of esterification (Karty, Chapter 21) and try to identify the rate-limiting step. Discuss the effect, if any, of the size (or bulk—what is the difference?) of the organic group on the rate constant for esterification. Where is substitution on the organic group most significant?

Be sure to analyze the results for each alcohol to see whether any differences you found are statistically significant.

The report will be due one week after the combined data are posted to the course web page.

> Please do not parrot back the discussion of mathematical rate laws given above. If you wish to discuss rate laws and reaction kinetics, you should review the material in your general chemistry textbook and cite it properly.

> No experimental section is required for this report!

Pre-laboratory assignment

Give at least two good reasons for using trifluoroacetic acid, rather than acetic acid, in this NMR experiment.

Synthetic experiments

The business of organic chemistry is building new molecular structures. These experiments will introduce some of the reactions in the synthetic organic chemist's toolkit, with which he builds the most intricate and beautiful structures imaginable.

An indispensable reference for these experiments is Zubrick's *Organic Chem Lab Survival Manual*.

Not all of these experiments will be used in any given academic year.

Considerations for Multistep Synthesis

It is rare that a useful compound may be synthesized in only one or two steps. Typically several intermediate compounds must be synthesized, purified and characterized along the way to the desired product.

Two issues become important in multi-step syntheses:

- Yield. A 60% yield may seem acceptable for an organic reaction. However, consider a procedure where five reactions must be carried out in sequence, each reaction using the product of the preceding step as starting material. To obtain the overall yield you must multiply the yield of each step; a 60% yield in each of five steps leads to an overall yield of 7.8%! The procedure is very inefficient. You could compensate by simply running the first step on a very large scale and accepting losses along the way, and in fact if the precursors are cheap this often is done in the "real world" – at least in academia, where getting to the final product is more important than efficiency or waste minimization. However, chemicals are usually not cheap, and it is even more expensive to dispose of large amounts of chemical waste. The better solution is to maximize the yield of each step, and to combine steps where possible. One of the most important parts of industrial research is process optimization, in which the object is to increase the overall yield and reduce the amount of waste generated from chemical manufacturing.

- Identity and purity of the intermediates. In a multi-step synthesis, it is essential that the identity of the product of each step be established. You do not want to proceed to the next step if you are not certain of the product from the previous step. Reasonable purity is also essential, as impurities often interfere with subsequent steps either by reacting preferentially with the reagents or simply diluting the reactants to uselessness. Note that small amounts of impurities are seldom a problem, and loss of product by recrystallization or some other purification step may actually do more harm than good by reducing the yield.

You can overcome some issues, both yield and the purification of intermediates, by combining as many steps as possible into a single process. For example, Grignard experiments combine two steps in the same reaction flask:

(a) Synthesis of the Grignard reagent, the organomagnesium halide compound.

(b) Reaction of the Grignard reagent with another compound.

Whenever you perform a multistep synthesis, report the yield of *each step* as well as the *overall yield*. That means the yield from your initial starting material. This should be similar to or identical to the result obtained by multiplying together the yields of the individual steps.

Saponification: the hydrolysis of fats to fatty acids

Based on a procedure developed by John Thompson, Lane Community College, and Michael Koscho, University of Oregon.

Vegetable oils and animal fats are "triglycerides," tri-esters of glycerine (1,2,3-propanetriol) with fatty acids. Fatty acids are long-chain carboxylic acid compounds, typically containing eight or more carbon atoms.

Soap is prepared by a reaction known as *saponification* (a word of Latin origin that means "making soap"). In a saponification reaction, an ester is heated with aqueous hydroxide to form the salt of the acid, along with the alcohol. For a triglyceride, the reaction with sodium hydroxide is

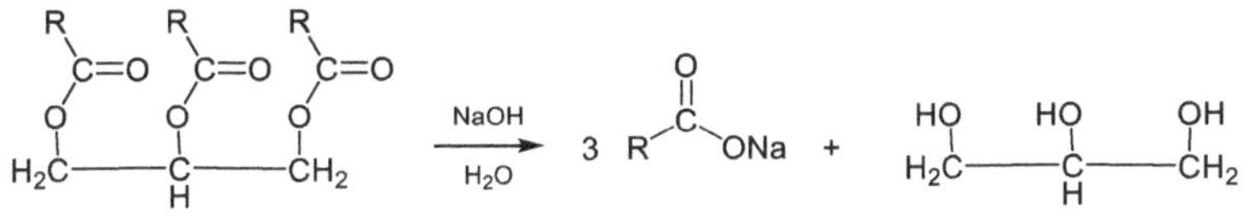

Reaction 1

In principle this reaction is reversible, but because of the high pH of the reaction conditions, all the carboxylic acids formed are in their unreactive sodium carboxylate forms.

There are a number of different fatty acids; almost all have even chain lengths, from butanoic acid (C_4) to stearic acid (C_{18}). They can also have carbon-carbon double bonds, which in biological fatty acids are always in the Z configuration. Fatty acids are sometimes referred to by their chain length and the number of double bonds, so that hexanoic acid is designated 6:0 while hex-3-enoic acid is 6:1. Some representative fatty acid structures are shown below, with their common names.

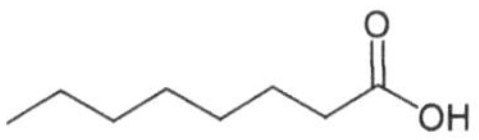

caprylic acid (8:0)

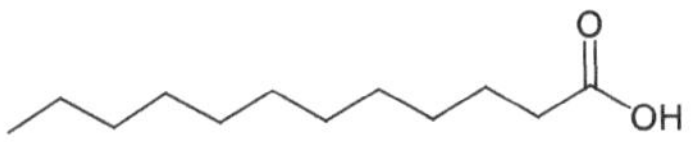

lauric acid (12:0)

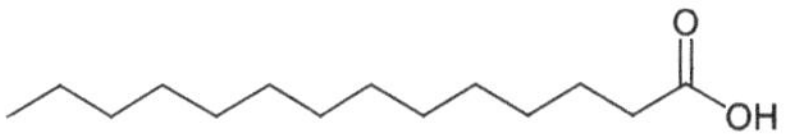

myristic acid (14:0)

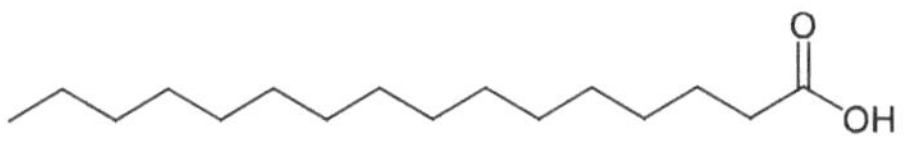

palmitic acid (16:0)

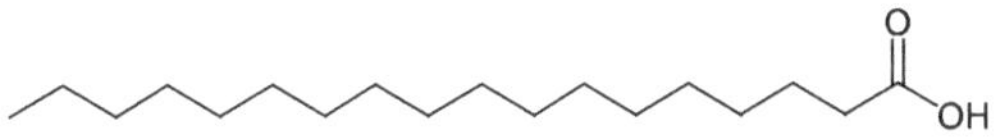

stearic acid (18:0)

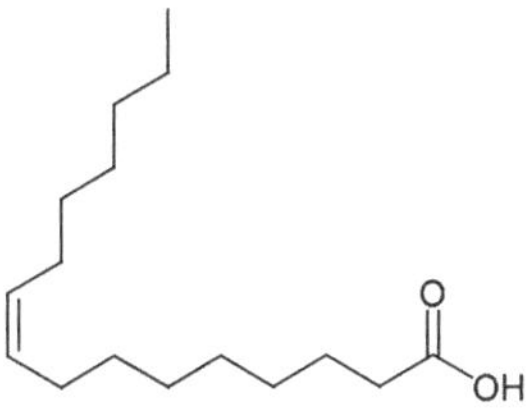

palmitoleic acid (16:1)

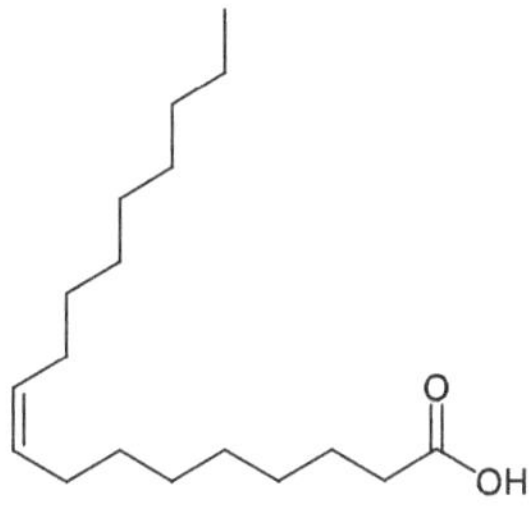

oleic acid (18:1)

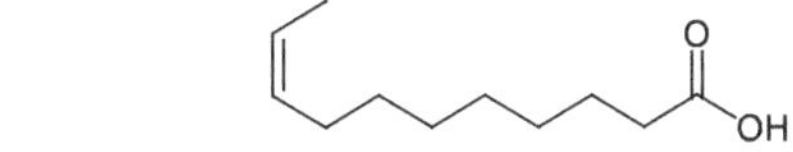

linoleic acid (18:2)

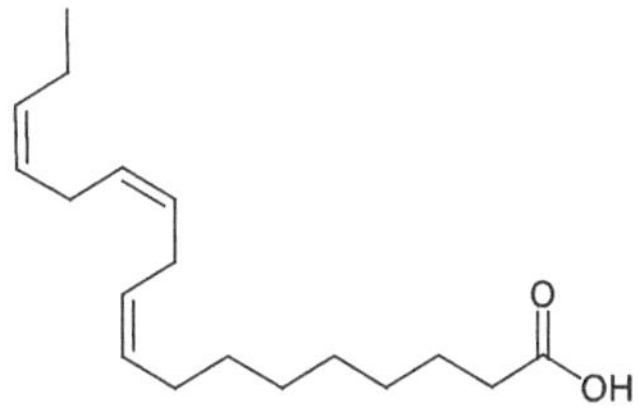

α-linolenic acid (18:3)

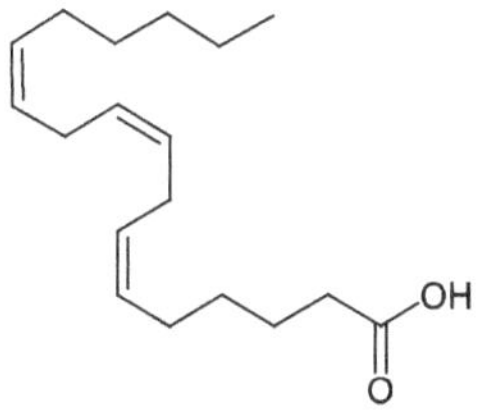

γ-linolenic acid (18:3)

Fatty acid compositions of a number of animal fats and vegetable oils are shown below. It can be seen that animal fats generally have a higher proportion of saturated fats than vegetable oils. This has consequences for whether a lipid is solid or liquid.

Average Fatty Acid Composition (by Percentage) of Selected Fats and Oils

	C_{10} C_8 C_6 C_4	C_{12} Lauric	C_{14} Myristic	C_{16} Palmitic	C_{18} Stearic	C_{20} C_{22} C_{24}	C_{16} Palmitoleic	C_{18} Oleic	C_{18} Ricinoleic	C_{18} Linoleic	C_{18} Linolenic	C_{18} Eleostearic	C_{20} C_{22} C_{24}
	Saturated Fatty Acids (No Double Bonds)						Unsaturated (1 Double Bond)			Unsaturated (>1 Double Bond) (2)	(3)	(3)	Unsaturated
Animal fats													
Tallow			2–3	24–32	14–32		1–3	35–48		2–4			
Butter	7–10	2–3	7–9	23–26	10–13		5	30–40		4–5			2
Lard			1–2	28–30	12–18		1–3	41–48		6–7			2
Animal oils													
Neat's foot				17–18	2–3			74–77					
Whale			4–5	11–18	2–4		13–18	33–38					17–31
Sardine			6–8	10–16	1–2		6–15	← 24–30 →					12–19
Vegetable oils													
Corn			0–2	7–11	3–4		0–2	43–49		34–42			
Olive			0–1	5–15	1–4		0–1	69–84		4–12			
Peanut				6–9	2–6	3–10	0–1	50–70		13–26			
Soybean			0–1	6–10	2–6			21–29		50–59	4–8		
Safflower				6–10	1–4			8–18		70–80	2–4		
Castor bean				0–1				0–9	80–92	3–7			
Cottonseed			0–2	19–24	1–2		0–2	23–33		40–48			
Linseed				4–7	2–5			9–38		3–43	25–58		
Coconut	10–22	45–51	17–20	4–10	1–5			2–10		0–2			
Palm			1–3	34–43	3–6			38–40		5–11			
Tung				← 2–6 →				4–16		0–1		74–91	

Z alkenes cannot pack as tightly in the solid state as can saturated chains; therefore, increasing the proportion of saturated fats increases the melting point. The melting points of some individual fatty acids are given in the table below:

saturated fatty acids	m.p. (°C)	unsaturated fatty acids	m.p. (°C)
12:0	44	16:1	−1
14:0	52	18:1	13
16:0	63	18:2	−9
18:0	70	18:3	−17
20:0	75	20:4	−50
22:0	81	20:5	−54

Because of this trend, soaps have traditionally been made using animal fats such as tallow or lard, which are high in saturated fatty acids. Traditionally, fat was heated with water and wood ashes, which are basic.

90

However, when lye (sodium hydroxide) became commercially available, people switched to the stronger base for soap making. Many soaps today are made from solid plant oils such as palm oil or coconut oil rather than animal fats.

In this experiment we will make soap using coconut oil and lye. We will take care to use slightly less lye than is needed to fully hydrolyze the coconut oil; we don't want excess lye in the soap as this would make it harshly caustic. The final product will be a mixture of soap (fatty acid sodium salts), glycerine, water and a small amount of monoglycerides (esters with just one fatty acid on glycerine). This soap will be safe for you to take home and use. You are encouraged to bring colognes or perfumes to mix with your soap if you wish.

Minimum safety standards for this experiment

1. You will be working with sodium hydroxide, as a pure solid and in concentrated solutions. This is a strong caustic. It will eat holes in your clothes and your skin if not washed off. When cleaning up, be sure to remove all jewelry and anything else covering skin that may have come in contact with sodium hydroxide.

2. Vegetable oil is flammable. Keep away from strong heat, sparks or flames.

3. The soap you produce should have a neutral pH but is still an irritant. You may use it for washing and bathing but should not use it near your eyes.

4. You should treat all waste liquids as caustics, as they will contain sodium hydroxide.

Disposal

All reagents, products and wastes may be disposed of by flushing them down the sink, if liquid, or placing them in the trash, if solid. Sodium hydroxide spills must be swept or wiped up, and the counter wiped down with a clean, damp paper towel.

Procedure

1. To a 250 mL beaker add about 50 g coconut oil, or about 45 g coconut oil and about 5 g of liquid vegetable oil. Add a large stir bar and heat to liquefy the coconut oil.

2. In a separate beaker, dissolve 8.5 g of sodium hydroxide in 24 mL of deionized water.

3. Carefully add the NaOH solution to the oil. Be sure your liquid oil is below the boiling point of water!

4. Heat at about 90-100°C until you have converted your mixture into molten soap, monitoring the temperature with a thermometer. You should notice a large increase in viscosity when the reaction is complete. You will need to heat above 90° for about an hour.

5. Transfer your molten soap into a mold and let cool.

6. Add a small piece of your soap to a test tube and add about 4-5 mL of deionized water. Cap the test tube with your thumb, shake it, and note what happens. Test the pH of your soap solution; it should be neutral.

7. Add a small amount of the supplied calcium chloride solution to your soap solution and note what happens.

For the report

Draw a complete mechanism for the saponification reaction; you may use "R" to represent the alkyl groups attached to the acid and alcohol groups.

Explain, in terms of chemical interactions, what you observed when you added calcium chloride to your soap.

Explain how soap is able to help dissolve non-polar dirt and grease.

Explain why monoglycerides, which form part of your product, are able to act in a similar way to soap.

glycerine monolaurate, a monoglyceride

Prelaboratory assignment

1. The average molecular weight of the triglycerides in coconut oil is 680 g/mol. How much NaOH, in grams, is required to completely saponify 50 grams of coconut oil? (Hint: how many moles of NaOH are required to saponify one mole of a triglyceride? See *Reaction 1*, above.)

2. How does this amount compare to the amount called for in the procedure?

3. Why is important to use **less** than the required amount of NaOH when making soap?

4. Are there any safety concerns that have not been adequately addressed in this procedure?

Dehydration of an alcohol

Alkenes may be synthesized by treating an alcohol with a strong acid. The acid protonates the hydroxyl group, converting it into the excellent leaving group, water. The elimination reaction proceeds by an E1 mechanism, with the protonated alcohol losing water and then being deprotonated by a base, creating a double bond and regenerating the acid catalyst. This reaction, the reverse of an *electrophilic addition* reaction, is discussed in Section 9.5 of Karty. The chemistry of the bromine test is discussed in Section 12.3.

In this experiment, cyclohexene will be synthesized from cyclohexanol. The alcohol is favored if the reaction is allowed to come to equilibrium; to drive the reaction forward, the alkene will be removed as it is formed by the technique of *distillation*. You will also be removing water from your product by using a *drying agent*.

Minimum Safety Standards for this experiment

1. Hot glass looks the same as cold glass! Before picking up a piece of glassware, be sure to check that it is cool enough to handle.

2. Reagents which have an odor or an appreciable vapor pressure may not be used outside the hood except in closed containers.

3. Look up the MSDS for each reagent used. More specific cautions and procedures are given below.

4. Cyclohexanol is water-soluble enough that residues may be cleaned with soap and water and flushed down the sink.

5. You will be using concentrated phosphoric acid. While this acid does not produce fumes, it is strongly corrosive! Be sure to wash your hands after mixing the initial reaction solution to remove any traces of phosphoric acid. Also wash after handling any other substance produced in this experiment.

6. Cyclohexene has a strong odor. Keep the product in the hood at *all times*. This includes contaminated glassware and paper towels. Rinse all contaminated glassware with acetone *in the hood*, into a **stoppered** waste container, before removing it from the hood. Waste contaminated with cyclohexene may be combined into the waste bottle provided.

7. Bromine is corrosive and toxic and must be used only in the hood. Be sure to familiarize yourself with the MSDS. Wash your hands after handling the bromine bottle. Be sure to ask for gloves before attempting to clean up spillage.

Disposal

aqueous waste	This will smell of cyclohexene and should be flushed down the sink in the hood. The residue in the still pot from the initial reaction and the aqueous solutions used for washing are both "aqueous waste."
acetone waste	Acetone rinsings should be collected in a large beaker and placed in the appropriate waste bottle.
bromine	Any waste containing bromine (it will be colored) should be put aside and given to the instructor for neutralization and disposal. This includes paper towels. Exception: the blank solution for your bromine test may be placed in the waste bottle provided.
cyclohexanol	Residues should be cleaned with *acetone* and flushed down the sink unless they are contaminated with cyclohexene.
cyclohexene	Residues should be cleaned with acetone in the hood and placed in the waste bottle provided.
drying agent	Drying agent should be placed in the container provided and will be disposed of by the instructor.
5% sodium carbonate	Anything that this has contacted must be rinsed with water.
phosphoric acid	Residues should be cleaned with *water* and flushed down the sink.

Procedure

Techniques used, from Zubrick, 9th Ed: jointware (Zubrick Chapter 4), clamps and joint clips (Zubrick, Chapter 18), distillation (Zubrick Chapter 19, Class 1: simple distillation), extraction and washing (Zubrick Chapter 15), drying an organic liquid (Zubrick Chapter 10)

1. Assemble a simple distillation apparatus using a 50-mL round-bottomed flask as your boiling pot and a 25-mL round-bottomed flask as your receiver. Be sure to clamp your apparatus properly.

2. Place 8 to 10 g cyclohexanol in your 50 mL flask. Add 3 mL concentrated phosphoric acid and a boiling chip or two. Attach the flask to the still head.

3. Heat your boiling flask until your mixture starts to distill. Continue to collect product, keeping the head temperature below 120°C. After the distillation is finished – you will have only a few mL of liquid left, and it will be dark yellow – transfer your product to a separatory funnel. It will form two layers: the *alkene* and *water* products of the reaction.

4. Rinse all of your distillation glassware with acetone (rinse the boiling flask with water first) in the hood, and allow it to dry while you perform the next steps.

5. Add about 5 mL of 5% sodium carbonate to the funnel. Mix well. Allow the phases to separate and remove the aqueous layer. Repeat the washing with a second portion of 5% sodium carbonate. Separate your product from the aqueous layer, being careful not to take any of the aqueous layer with your product.

6. In a 25-mL Erlenmeyer flask, add sufficient anhydrous calcium chloride to the product to remove the remaining water. Let it stand for five or ten minutes. If the drying agent appears to liquefy, add more of it.

7. To redistill your product, you should have at least 5 mL of crude product. You can still get a yield and a bromine test on the crude product, even if you don't redistill.

8. Reassemble your still, using your clean, dry 25 mL flask as the boiling pot and collecting in any clean, dry flask; be sure to tare your receiving flask, with its stopper! Don't forget a fresh boiling chip.

9. Distill the product. Do not let the head temperature rise above 90°. Continue boiling until the head temperature begins to fall, or until only a few drops are left in the boiling flask.

   Never let a boiling flask go completely dry, as it will overheat and could shatter!

10. Weigh the final product and report the yield. After weighing, bring the product to the instructor, who will perform a bromine test according to the procedure described below.

11. Bromine has historically been used to test for the presence of carbon-carbon multiple bonds. Place a sample of your product (about 0.5 mL) and a sample of cyclohexanol in two test tubes. Now add a few drops of bromine solution to each test tube. Describe the bromine and your test solutions before they are mixed. Describe what happens after mixing.

Prelaboratory questions

1. Why do we distill the alkene away from the reaction mixture, rather than just allowing the reaction to run at room temperature?

2. Describe how you will be able to determine, by adding bromine (Br_2), whether your product is an alkene.

3. What is the purpose of each of the following reagents: phosphoric acid? sodium carbonate?

4. Are there any extraordinary hazards present in this experiment? Defend your answer. Consider not only the chemicals used, but also their amounts and the conditions.

For the report…

Draw the mechanism of the dehydration reaction, including curved arrows. Discuss the chemistry of the bromine test and draw a mechanism. Report your yield correctly.

Reduction of a Ketone: Benzil

Ketones, which may be synthesized from alcohols by oxidation, may also be converted into alcohols by reduction with sodium borohydride.

In this experiment, we will reduce *benzil* (1,2-diphenyl-1,2-ethanedione) to *hydrobenzoin* (1,2-diphenyl-1,2-ethanediol); the incomplete reduction product is *benzoin* (2-hydroxy-1,2-diphenylethanone). Observe that multiple stereoisomers of the product are possible. We will examine the degree of conversion by TLC.[1]

benzil   benzoin   hydrobenzoin

Techniques used from Zubrick 9th Ed: Crystallization and vacuum filtration (Chapter 13); thin-layer chromatography (Chapter 27).

Minimum Safety Standards for this experiment

1. Benzil is a mild irritant. Wash after handling.

2. Benzoin is an irritant. Wash after handling.

3. Hydrobenzoin is considered non-toxic. Wash after handling.

4. Reagent alcohol is an irritant. It is volatile, and so one is susceptible to breathing the vapor; it may also be absorbed through the skin. It should be used in the hood.

5. Sodium borohydride is a caustic powder; it can damage mucous membranes and other exposed, moist areas of the body. Handle with care and wash thoroughly after using. Wash spatulas after they are exposed to this substance, to keep them from corroding.

6. Dichloromethane is relatively innocuous. It is a mild mutagen and mildly toxic. Treat it with respect, handle *only* in the hood, and wash your hands after handling.

7. Ethyl acetate is an irritant, but otherwise safe in the quantities we are using. Treat it with respect, handle *only* in the hood, and wash your hands after handling.

Disposal

Benzil	Residues should be swept into the trash.
Benzoin	Residues should be swept into the trash.
Hydrobenzoin	Residues should be swept into the trash. Your solid product may be disposed of in the trash.

---

[1] This procedure was published online in 2010 by Dr. M. Milkevitch, Philadelphia University, Philadelphia, PA.

Dichloromethane	Wipe up spillage with paper towels and allow the dichloromethane to evaporate in the hood. The paper towels may then be disposed of in the wastepaper basket. Developing solvent should be placed in the organic waste bottle; be sure to log your deposits.
Alcohol	Flush down the sink. Wipe up spillage with paper towels and allow the alcohol to evaporate in the hood before disposing of the paper towels in the trash.
Ethyl acetate	Wipe up spillage with paper towels and allow the ethyl acetate to evaporate in the hood. The paper towels may then be disposed of in the wastepaper basket. Developing solvent and spotting solutions should be placed in the organic waste bottle.
Filter pipet	May be disposed of in glass waste.
Sodium borohydride	Water/ethanol solutions may be flushed down the sink. Spillage should be swept into the sink, then flushed with plenty of water.

Procedure

1. Weigh out 200 mg of benzil into a 25 ml erlenmeyer flask. Add 2 ml of reagent alcohol, and swirl to mix.

2. Add 75 mg of sodium borohydride and swirl to mix. Let the mixture stand 10 minutes with occasional swirling. The yellow color should disappear as the benzil is reduced.

3. After this, **slowly** add 2 ml of deionized water. Carefully heat the mixture to boiling; the solution should be clear and colorless. Your mixture will bubble vigorously as the remaining sodium borohydride reacts with water. Continue heating gently until the fizzing subsides; be careful not to boil away your solvent.

4. If you have visible solid, add alcohol a few drops at a time until the solid dissolves.

5. Add an additional 2 ml of hot, deionized water. Allow the solution to cool undisturbed until it has crystallized at room temperature for some time. Cool your mixture in an ice bath for a few minutes before vacuum-filtering; wash with minimal <u>cold</u> water.

6. Place your product in a dish for drying. Dissolve a few milligrams of your product in ethyl acetate for TLC. As standards, you should also prepare solutions of a few milligrams each of *meso*-hydrobenzoin[1] and benzoin. Run all three solutions on the same TLC plate, developing in 9:1 dichloromethane/ethyl acetate.

7. One student's product will be selected on the basis of TLC results, to use for an IR spectrum. We will discuss the IR spectrum as a lab group, comparing it with authentic spectra.

8. Obtain a melting point of your product after it has dried for a few days.

---

[1] 1,2-diphenyl-1,2-ethanediol

Prelaboratory questions

1. What is the purpose of the ethanol? Why don't we just use water?

2. Why is sodium borohydride relatively stable in the presence of base, but not in the presence of acid? Explain the chemistry involved.

3. How many stereoisomers of benzil are there? Draw them. What are their melting points?

4. How many stereoisomers of benzoin are there? Draw them. What are their melting points?

5. How many stereoisomers of hydrobenzoin are there? Draw them. What are their melting points?

6. Are there any extraordinary hazards present in this experiment? Defend your answer. Consider not only the chemicals used, but also their amounts and the conditions.

For the report

Did you reduce all of your benzil? Did you reduce it to benzoin or to hydrobenzoin, or both?

Draw a reaction mechanism that explains the formation of your product.

Melting points of hydrobenzoin

	m.p.	source
meso-hydrobenzoin	137-139°	Aldrich catalog
(±)-hydrobenzoin	118-123°	Alfa Aesar catalog
(R,R)-hydrobenzoin	146-149°	Aldrich catalog
(S,S)-hydrobenzoin	148-150°	Aldrich catalog

Reductive amination of vanillin

Based on Touchette, K.M. *J. Chem. Educ.* **2006**, *83*, 929-930

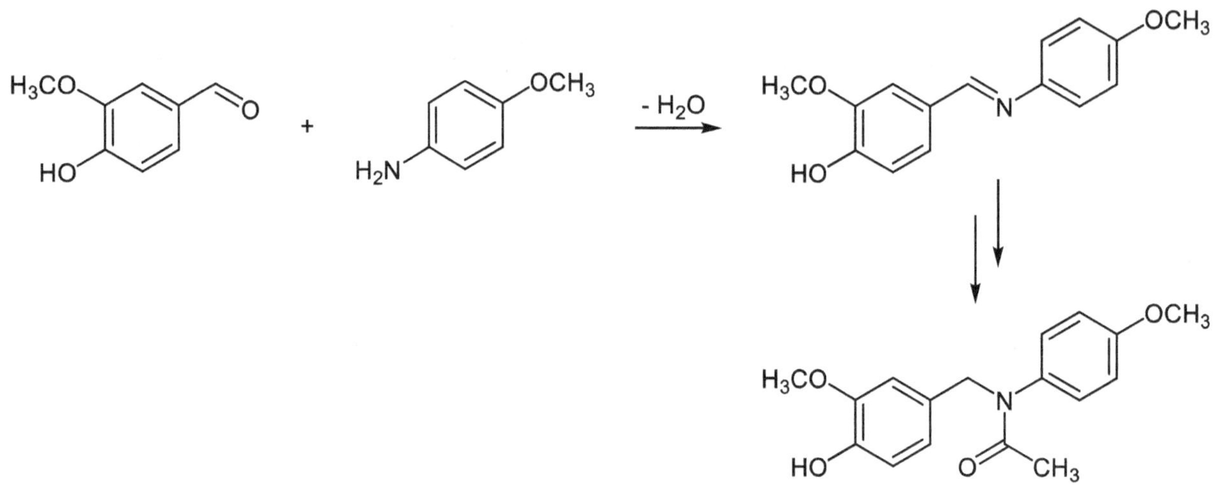

Reductive amination is normally a one-pot procedure in which an aldehyde or ketone reacts with a primary amine, forming an imine that is selectively reduced *as it is formed* by using a weakened reducing agent that will not easily react with C=O bonds but *will* react with C=N bonds. This requires more exotic reducing agents such as hydrogen gas over nickel, or weakened hydride donors such as $NaBH_3CN$.

In this experiment, we will react vanillin with *p*-anisidine to generate an imine that will be reduced with sodium borohydride, then acetylated with acetic anhydride. The reactions we will use will be green: they will use minimal organic solvent, and no solvent at all for the first step.

Minimum Safety Standards for this experiment

1. Vanillin is innocuous. Clean up spills immediately.

2. *p*-Anisidine is only mildly toxic, but is a severe skin irritant and a suspected carcinogen. Handle with care, clean up spills immediately and wash your hands thoroughly after handling.

3. Reagent alcohol is flammable and mildly toxic. It is best used in the hood.

4. Sodium borohydride is corrosive/caustic and will cause burns if left on the skin. Wash your hands after handling. **Immediately wash any metal tools** used to handle sodium borohydride **with water**.

5. Glacial acetic acid is corrosive and will cause burns if left on the skin or if the vapor is inhaled. Wash your hands after handling and use only in the hood.

6. Acetic anhydride is corrosive and will cause burns if left on the skin or if the vapor is inhaled. Wash your hands after handling and use only in the hood.

7. Hexanes are flammable and moderately neurotoxic. Do not inhale the vapors; use only in the hood.

Disposal

vanillin	Sweep up spills and place in the trash.
p-anisidine	Sweep up spills and place in the trash.
reagent alcohol	Flush down the sink. Residues and spills may be cleaned with water or allowed to evaporate in the hood.
sodium borohydride	All residues should be flushed down the sink. Small spills should be swept up and flushed down the sink.
glacial acetic acid	Flush down the sink in the hood. Wipe up spills and rinse the towel with plenty of water before discarding.
acetic anhydride	Flush down the sink in the hood. Wipe up spills and rinse the towel with plenty of water before discarding.
hexanes	Leftover amounts larger than 0.5 mL must be placed in a waste bottle. Residues on glassware may be cleaned with soap and water. Spills should be wiped up and allowed to evaporate before disposing of the paper towels in the trash.

Procedure

1. Measure out 5 mmol each, of vanillin and of *p*-anisidine. In a 100-mL beaker, crush and mix the reagents with a heavy stirring rod for 6-10 minutes until a uniform powder is formed; note any color and texture changes. You will periodically need to scrape material off the walls of the beaker with a spatula.

2. Add 15 mL of reagent alcohol to the beaker and stir. Again, be sure to scrape off any material adhering to the walls of the beaker.

3. While stirring, carefully add about 0.1 g of sodium borohydride in two or three portions. Continue stirring and scraping until all solids dissolve. Allow to stand for a minute or two.

4. Carefully add about 2 mL of glacial acetic acid to destroy any remaining borohydride.

5. While stirring, add about 2 mL of acetic anhydride. Warm the mixture in a water bath (50-70°) for ten minutes, stirring occasionally.

6. Transfer the mixture to a 250-mL beaker and add 100 mL of cold water. Allow to crystallize for at least ten minutes. Collect the solid by vacuum filtration.

7. If there is time, recrystallize some of your product from hexanes. Keep enough of your crude product to obtain a melting point.

8. Obtain IR and NMR spectra of your product, as well as a melting point.[1] Is your product what you expected? Is it pure? If you recrystallized, is the product before recrystallization pure?

---

[1] Melting ranges that are less than two degrees indicate that a substance is pure, even if you don't know the expected melting point.

This product will not dissolve well in chloroform. You should instead use DMSO-d_6. Do not try to put TMS in your NMR solutions! It will not dissolve well in DMSO.

While the product will dissolve in acetone-d_6, that solvent is not recommended because it will exchange hydrogens with the phenolic OH group, and also bury one of your peaks.

For the report

Draw the mechanism of each step in the reaction sequence for your report. Use melting point and spectral data to characterize your product.

Calculate the theoretical atom economy for this synthesis. Include all reagents that provide any portion of the product. The three balanced reactions you should use are shown below.

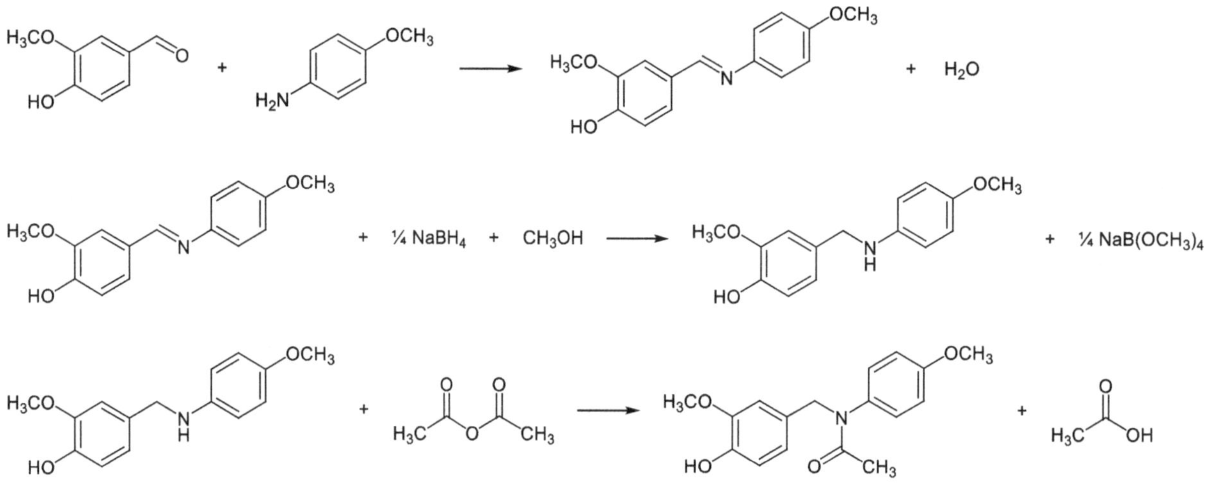

Report the Green Metrics for this experiment, as listed in the "Green Metrics" handout. Discuss how "green" this experiment turned out to be, based on your calculated metrics.

Pre-laboratory questions

1. Calculate the mass of 5 mmol of vanillin and 5 mmol of p-anisidine.

2. Draw the mechanism of the first step of the synthetic sequence shown above, in which an amine reacts with an aldehyde to form an imine.

3. What peaks do you expect to see in the ^{1}H NMR spectrum of the final product? Include integration as well as estimates of the chemical shift.

4. Discuss any extraordinary hazards in this experiment, paying particular attention to things you think are not sufficiently emphasized in this lab procedure.

Synthesis of an Ester: Benzoin Acetate

This reaction is an esterification using an alcohol and an acid anhydride. The reaction takes place by nucleophilic acyl substitution of the alcohol on the anhydride.[1] What is the nucleophile in this reaction? The leaving group?

Techniques used from Zubrick, 9th Ed: addition and reflux (Chapter 22); using a hot water bath (not in Zubrick, but see also Chapter 17); extraction and washing an organic liquid (Chapter 15); drying an organic liquid (Chapters 9 and 10); crystallization, vacuum filtration and washing a solid product (Chapter 13).

Minimum Safety Standards for this experiment

1. Hot glass looks the same as cold glass! Before picking up a piece of glassware, be sure to check that it is cool enough to handle.

2. Acetic anhydride is moderately corrosive and smelly; it combines with water (including the water in your tissues) to produce acetic acid in concentrated form. Handle it only in the hood, and wash your hands after handling.

3. Benzoin is somewhat toxic, but innocuous in the quantities we will use. Wash your hands after handling.

4. Be sure to wash your hands after mixing the initial reaction solution to remove any traces of acetic anhydride. Wash after handling any other substance produced in this experiment.

5. You will be boiling water during this experiment; be cautious and avoid steam burns.

6. Methanol is toxic in moderate quantities. However, as long as you use it in the hood and avoid contamination you should be perfectly safe.

Disposal

Acetic Anhydride	Residues should be cleaned with water and flushed down the sink in the hood.
Benzoin	Minor spillage should be swept up and placed in the waste basket.
Benzoin Acetate	Small amounts may be cleaned with acetone and flushed down the sink. Product residue should be placed in the waste container provided.
Dichloromethane	Most of the dichloromethane will be evaporated. Wipe up spillage with paper towels and allow the dichloromethane to evaporate in the hood. The paper towels may then be disposed of in the wastepaper basket.
Methanol	Residues may be cleaned with water and flushed down the sink.

---

[1] See Chapter 21 of Loudon.

Procedure

1. Place 2 g benzoin in a 50-mL round-bottomed flask with a magnetic stirring bar. Attach the condenser to the flask. Circulate water through the condenser.

2. Add 10 mL acetic anhydride through the top of the condenser, and stir in a boiling water bath for 30 minutes.

3. Remove from the boiling water bath and pour the mixture into an Erlenmeyer flask. Add 20 mL water. What change do you observe? Cool to room temperature.

4. If crystals have not yet formed, add about 10 mL of ice-cold water. Wait for crystals to form.

5. After cooling the product 5 minutes in an ice/water bath, vacuum-filter the product and wash it with three 1-ml portions of *ice-cold* methanol. Allow the product to dry for at least 24 hours (dry to constant weight).

6. Determine the mass and melting point of the product. The melting point of pure benzoin acetate is 83°C.[1] What is the melting point of benzoin?

For the report

Report the yield of this reaction. Discuss the spectra and all reactions performed in this experiment in your report. How would you most easily determine the presence of unreacted benzoin, assuming you had enough product?

Pre-laboratory assignment

1. Compare the mechanism of an S_N2 reaction (Chapter 9) with the mechanism of nucleophilic acyl substitution (the Fisher esterification reaction, Chapter 20). How are they similar? How are they different?

2. The reaction of acetic anhydride with simple alcohols like methanol is rapid, exothermic and occurs at room temperature. Explain why this reaction requires a boiling water bath.

3. Are there any extraordinary hazards present in this experiment? Defend your answer. Consider not only the chemicals used, but also their amounts and the conditions.

---

[1] R.C. Weast and M.J. Astle, eds., *Handbook of Chemistry and Physics, 61st Ed.*, Boca Raton: CRC Press, Inc. (1980).

Esterification of benzoic acid[1]

In this experiment, we use the Fischer Esterification reaction to convert benzoic acid into its methyl ester. The technique is simple: we reflux the acid in methanol in the presence of catalytic sulfuric acid. The mechanism of this reaction is described in Karty, Chapters 20 and 21.

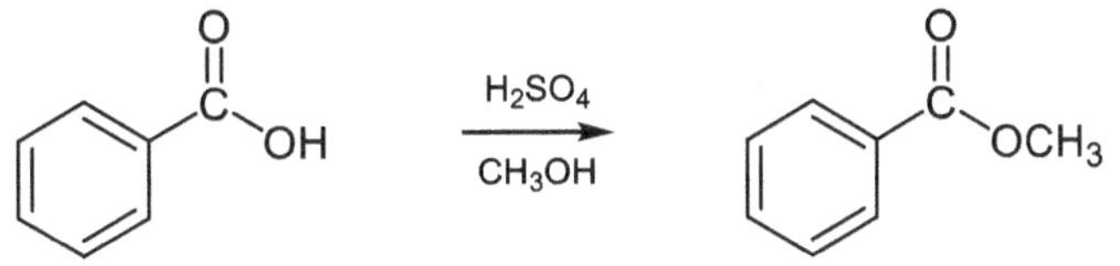

Techniques used from Zubrick, 9th Ed: simple distillation (Chapter 19), extraction and washing (Chapter 15), reflux and addition under water-free conditions (Chapter 22)

Minimum safety standards for this experiment

1. Hot glass looks the same as cold glass! Before picking up a piece of glassware, be sure to check that it is cool enough to handle.

2. Reagents which have an odor or an appreciable vapor pressure may not be used outside the hood except in closed containers.

3. Look up the MSDS for each reagent used. More specific cautions and procedures are given below.

4. Concentrated sulfuric acid causes severe burns. Be careful not to spill it on yourself! You may not notice the burn until you wash your hands!

Disposal

aqueous layers from extractions	Flush down the *hood* sink.
benzoic acid	Dry spillage may be swept up and thrown in the trash; aqueous or methanolic solutions may be flushed down the sink.
drying agent	Place on a paper towel in the hood until it no longer smells, then place in the wastebasket.
ether	Place ether residues in the waste bottle provided; small amounts may be poured onto the benchtop in the hood and allowed to evaporate. Rinse containers with acetone or isopropyl alcohol and flush the rinsings down the *hood* sinks.
methanol	Flush down the sink.
methyl benzoate	Place in the storage bottle provided. All glassware that is contaminated with methyl benzoate must be rinsed into the sink *in the hood*, then washed with soap and water.

---

[1] This procedure is taken from Williamson, K.L. Macroscale and Microscale Organic Experiments, 3rd ed.; Houghton Mifflin: Boston, 1999; pp. 480-1.

Procedure

1. In a 50-mL round-bottomed flask, combine 6 g benzoic acid and 20 mL methanol. *Carefully* add 2 mL of concentrated sulfuric acid. Swirl to mix.

2. Add a boiling chip, attach a reflux condenser, and heat to boiling. Boil the mixture gently for 30 minutes.

3. Cool the solution and decant into a separatory funnel containing 50 mL of water. Rinse the flask with 20-25 mL of ether and add the ether to the funnel. Separate. Wash the organic layer with a 25-mL portion of water, followed by a 25-mL portion of 5% sodium bicarbonate,[1] and finally a 25-mL portion of saturated aqueous sodium chloride.

4. Dry the organic layer. Decant the dry ether into a round-bottomed flask and wash the drying agent with another 5-10 mL of ether. Distill the mixture at atmospheric pressure; your product is the material boiling above 190°.[2]

5. Analyze your product by IR spectroscopy. How do you know it's an ester? How do you know it's dry? Record the yield for this step in your lab notebook.

For the report

Draw the mechanism of the esterification reaction. Determine the yield.

Pre-laboratory questions

1. What is the function of the sulfuric acid?

2. What hazards are associated with ether? Could you substitute another solvent for the ether during the extraction? Suggest a possible substitution.

3. How will you confirm the identity of your product?

4. Are there any extraordinary hazards present in this experiment? Defend your answer. Consider not only the chemicals used, but also their amounts and the conditions.

---

[1] Shake gently (but not too gently) until the funnel no longer "burps."

[2] Do not run water through the condenser for the latter part of the distillation; running water through a condenser while still temperatures are above about 120° can crack the glass.

Grignard synthesis of triphenylmethanol[1]

The Grignard reaction is a two-step, one-pot reaction.

In the first step, an organic halide – in this case, bromobenzene – is combined with magnesium to form an organomagnesium halide or *Grignard reagent*.

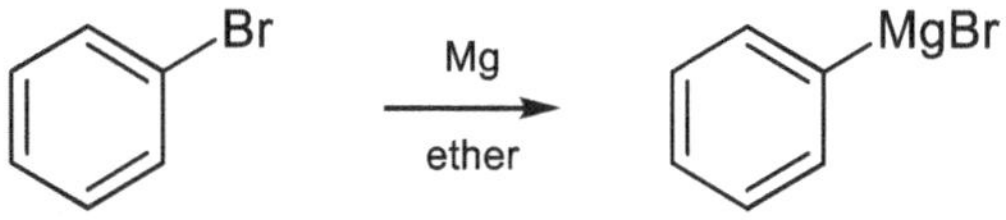

The mechanism by which this happens is a complex series of electron-transfer, dissociation, and coupling reactions, and is beyond the scope of this course.

The Grignard reagent is then reacted with some other compound. In this case, we are reacting it with an ester, methyl benzoate. The Grignard displaces any leaving group and also adds to any carbonyl group; see Karty, Chapter 20, for the detailed mechanism.

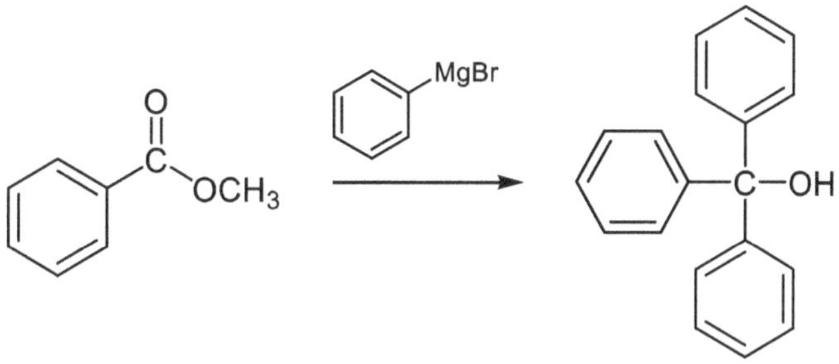

Techniques used from Zubrick, 9th Ed: extraction and washing (Chapter 15), reflux and addition under water-free conditions (Chapter 22)

All quantities in this experiment must be calculated, by you, according to the following criteria:

- You will use no more than 3 grams of methyl benzoate. Measure methyl benzoate by volume, not by weight!

- You should have a *ca.* 2-5% excess of Grignard over methyl benzoate (that is, for every mole of methyl benzoate you have, you should make 2.04-2.10 moles of phenylmagnesium bromide since the stoichiometry is 1:2).

- You should have a *ca.* 1-2% excess of bromobenzene over magnesium when you make the Grignard reagent. Measure the bromobenzene by volume, not by weight!

Never fear, the pre-lab walks you through all these calculations.

---

[1] This procedure is taken from Williamson, K.L. Macroscale and Microscale Organic Experiments, 3rd ed.; Houghton Mifflin: Boston, 1999; pp. 457-61.

Minimum safety standards for this experiment

1. Hot glass looks the same as cold glass! Before picking up a piece of glassware, be sure to check that it is cool enough to handle.

2. Reagents which have an odor or an appreciable vapor pressure may not be used outside the hood except in closed containers.

3. Look up the MSDS for each reagent used. More specific cautions and procedures are given below.

4. Your ether solution of Grignard reagent is pyrophoric and must be treated with great respect.

5. 10% sulfuric acid is not as bad as the concentrated acid you used last week. Nevertheless, treat it with respect!

Disposal

aqueous layers from extractions	Flush down the *hood* sink.
bromobenzene	Small quantities (less than 0.5 mL) may be flushed down the sink, rinsing with acetone or isopropyl alcohol.
drying agent	Place on a paper towel in the hood until it no longer smells, then place in the wastebasket.
ether	Place ether residues in the waste bottle provided; small amounts (less than 5 mL) may be poured onto the benchtop in the hood and allowed to evaporate. Rinse containers with acetone or isopropyl alcohol and flush the rinsings down the *hood* sinks.
magnesium	Place spillage in the trash.
methyl benzoate	Place any leftover methyl benzoate in the storage bottle provided. All glassware that is contaminated with methyl benzoate must be rinsed into the sink *in the hood*, then washed with soap and water. Rinse your glassware free of soap!
mother liquor from crystallization	Place in the waste bottle provided.
triphenylmethanol	Final product, once all analysis is complete, should be placed in the storage bottle provided.

Procedure

1. **Your equipment is in the drying oven!** This includes a 100-mL round-bottomed flask, a separatory funnel with the Teflon stopcock removed, a condenser, and a Claisen adapter.

2. Begin by assembling a drying tube, and weighing out the appropriate amount of magnesium. Fit your drying tube to your thermometer adapter.

3. Remove your glassware from the oven, place your magnesium in your round-bottomed flask and assemble an apparatus for reflux and addition. The Claisen adapter goes into the top of the RB flask, and the condenser and separatory funnel into the Claisen adapter. The drying tube goes into the top of

the condenser. You should put a stopper into the top of the separatory funnel, or another drying tube if materials are available. **Glassware should be assembled before it cools to room temperature.**

4. Mix the appropriate amount of bromobenzene – measured by volume, not by mass! – with about 10 mL of dry ether[1] in the separatory funnel and run enough of the mixture into the flask to immerse the magnesium. The instructor will start your reaction by breaking several pieces of magnesium in the flask.[2] The reaction should start within a few minutes (you will see bubbles, and the ether will turn metallic grey or brown).

5. If the reaction does not begin to boil on its own within five or ten minutes, bring the reaction mixture to a **gentle** boil (only a few bubbles forming!) Once the reaction begins, be ready to remove heat if it gets too vigorous!

6. Once the reaction starts, keep it going by swirling when the boiling action slows down. If swirling does not increase the boiling rate, add the about half of the remaining mixture in the addition funnel and swirl again. Repeat, for the last portion of the mixture in the funnel.

7. After you have added the last of the bromobenzene mixture, put another 10-15 mL of dry ether in the separatory funnel. Add ether, 2-4 mL at a time, whenever swirling is no longer effective in keeping the boil going.[3] When all ether has been added and the reaction slows down (as it will do near the end) you should apply gentle heat so that the mixture continues to boil. You should continue to swirl the reaction periodically to mix it. The reaction is complete when only a few small pieces of metal or metal contaminants remain in the flask.

8. Mix the appropriate amount of methyl benzoate – measured by volume, not by mass! – with about 10 mL of dry ether in the separatory funnel. Cool the reaction flask briefly in an ice-water bath, then remove the bath. Add the methyl benzoate solution, 2-3 mL at a time, swirling to mix and using the ice bath as needed to control the reaction. After addition is complete, reflux the mixture for 20-25 minutes (gently! When this reaction starts, it's exothermic at first!)

9. Pour the reaction mixture into a 250-mL Erlenmeyer flask containing about 50 mL of 10% sulfuric acid and about 25 g ice.[4] Rinse the reaction flask by vigorously swirling a combination of about 2 mL of 10% sulfuric acid and about 20 mL of ordinary ether, being sure to break up the solids in the round-bottomed flask with a spatula during the rinse. Swirl the mixture in the Erlenmeyer flask until the solids have dissolved;[5] you may need to add either more ether or more 10% sulfuric acid.

10. Separate the layers and wash the organic layer with a ~25-mL portion of 10% sulfuric acid, then with a similar portion of saturated sodium chloride solution. Dry the ether layer, and decant it into a clean, dry 125-mL Erlenmeyer flask (provided to you). Wash the drying agent with a small amount of ether and add the ether to the flask.

---

[1] You may want to rinse your graduated cylinder(s) with dry ether.

[2] This is to expose a fresh magnesium surface; the reaction takes place at the surface of the magnesium.

[3] Ether is necessary to keep the Grignard reagent in solution. If there is not enough ether, the Grignard reagent accumulates on the magnesium surface and prevents the reaction between bromobenzene and magnesium from happening.

[4] Don't waste your time trying to get precisely 25 grams! Time is precious in this experiment.

[5] You will see residual magnesium "fizzing" as it reacts with the acid to generate hydrogen gas.

At this point the Erlenmeyer flask should be **tightly** stoppered and set aside until the following week. The flask must, of course, be properly labeled.[1]

Be sure to clean your glassware and put it in the drying oven for tomorrow's lab section. If you took it out of the oven, it goes back in the oven. Of course, if there is no "tomorrow's lab section" this does not apply!

One week later...

11. Add about 25 mL of heptane to the flask, then heat gently on a hot plate until you begin to see crystals of triphenylmethanol.[2] Remove the flask from the heat and allow crystals to form, first at room temperature and then at 0°. Filter the crystals, washing with heptane or petroleum ether, and determine the yield from your first crop. Discard the mother liquor in the waste bottle provided.

    After a few minutes of air drying, use TLC to check your product against biphenyl, the primary by-product. A developing solvent will be provided for you; both compounds dissolve well in acetone.

12. Allow your product to dry for a day or two. Analyze your product by melting point. How pure do you think it is?

Pre-laboratory questions

1. What is the most important hazard associated with this experiment? How can you minimize it?

2. How much methyl benzoate did you obtain last week? (Do not enter more than 3 grams.) This is the amount you will use this week.

   _____ grams methyl benzoate

   How many **moles** of methyl benzoate are in that many grams?

   _____ moles methyl benzoate

   What is the **volume** of methyl benzoate you will use?

   _____ mL methyl benzoate

   How many **moles** of Grignard will you need to react with your methyl benzoate? (Consider reaction stoichiometry!) Add 5% to this number and enter it in the blank; this is the number of moles of Grignard you will be making.

   _____ moles Grignard

   How many **grams** of magnesium will you need to make that many moles of Grignard?

   _____ grams magnesium

   How many **grams** of bromobenzene will you need? Add 2% to this number and enter it in the blank; this is the number of grams of bromobenzene you will use.

   _____ grams bromobenzene

---

[1] If crystals form while the ether solution is standing, you may not want to add heptane in step 11!

[2] The purpose of this is to evaporate enough ether that the triphenylmethanol will crystallize.

How many milliliters of bromobenzene are there in that many grams? _____ mL bromobenzene

3. How will you confirm the identity of your final (triphenylmethanol) product?

4. Are there any extraordinary hazards present in this experiment? Defend your answer. Consider not only the chemicals used, but also their amounts and the conditions.

For the report

Draw the mechanism of the reaction of the Grignard with your ester.

- Formation of the Grignard from magnesium and bromobenzene is not a standard mechanism and is actually pretty complex, so you need not draw it.

Determine the yield.

The Grignard reaction: synthesis of dyes

Modification of Taber, D.F.; Meagley, R.P.; Supplee, D. *J. Chem. Educ.* **1996**, *73*, 259-260

Introduction

The addition of a Grignard reagent to a carbonyl compound is a common carbon-carbon bond forming reaction. Addition to a ketone or aldehyde results in conversion of the carbonyl group into an alcohol; but in an ester, the carbonyl group can reform by losing the alkoxide leaving group, then add another mole of the Grignard reagent.

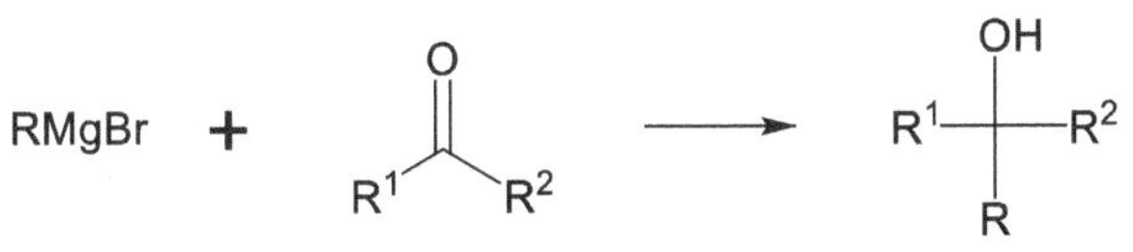

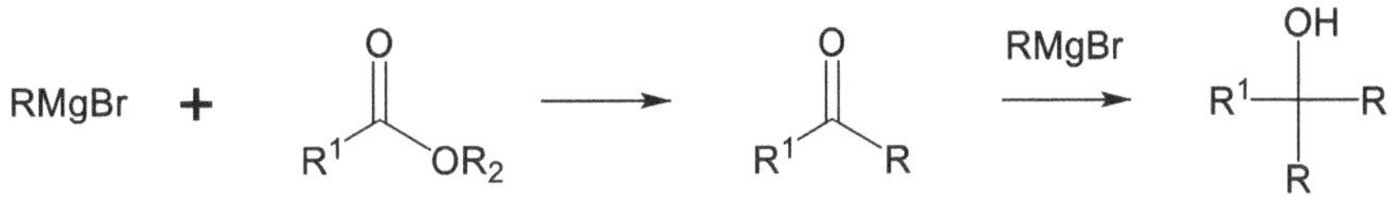

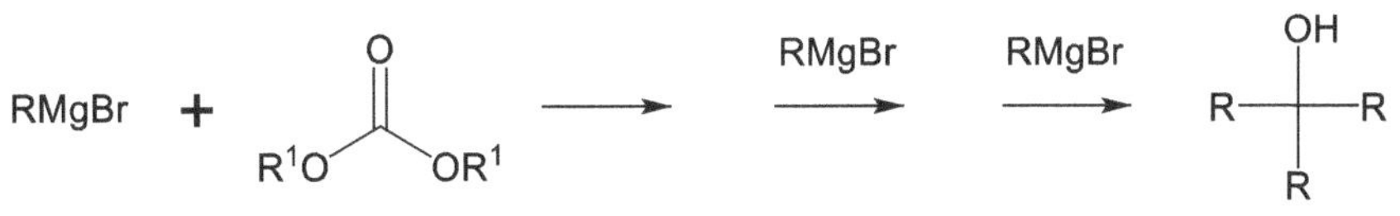

In this reaction, we will use 4-(dimethylamino)phenylmagnesium bromide as the nucleophile to synthesize two highly-conjugated, cationic dyes: malachite green and crystal violet. The only difference is the starting material: methyl benzoate or diethyl carbonate.

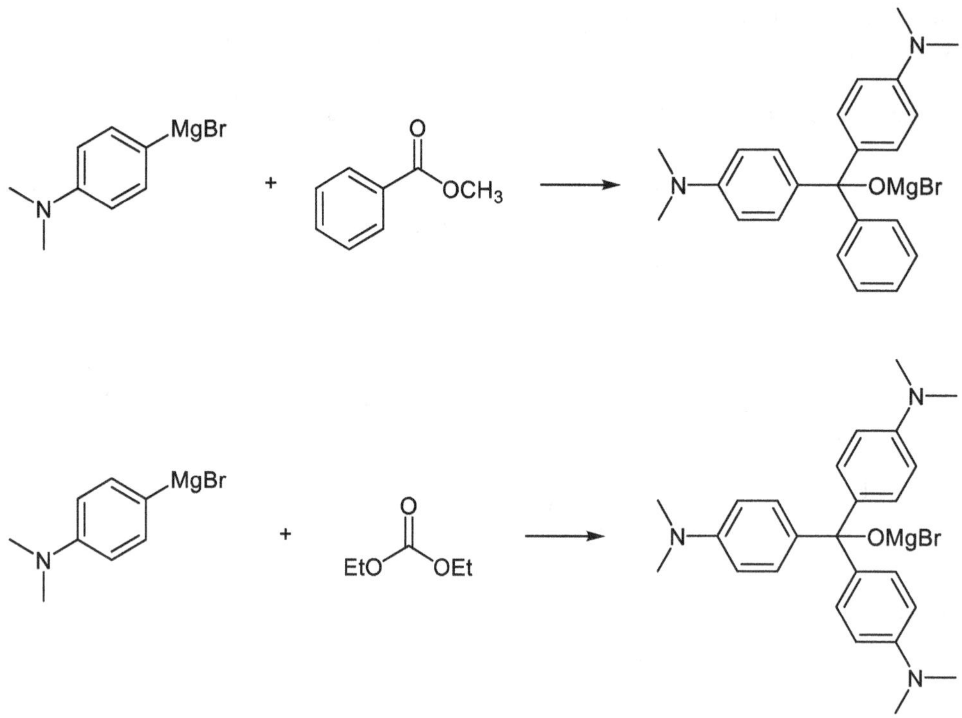

After the Grignard reaction, 10% HCl is used to form triarylcarbonium chloride compounds, strongly resonance-stabilized by electron donation from the dimethylamino groups.

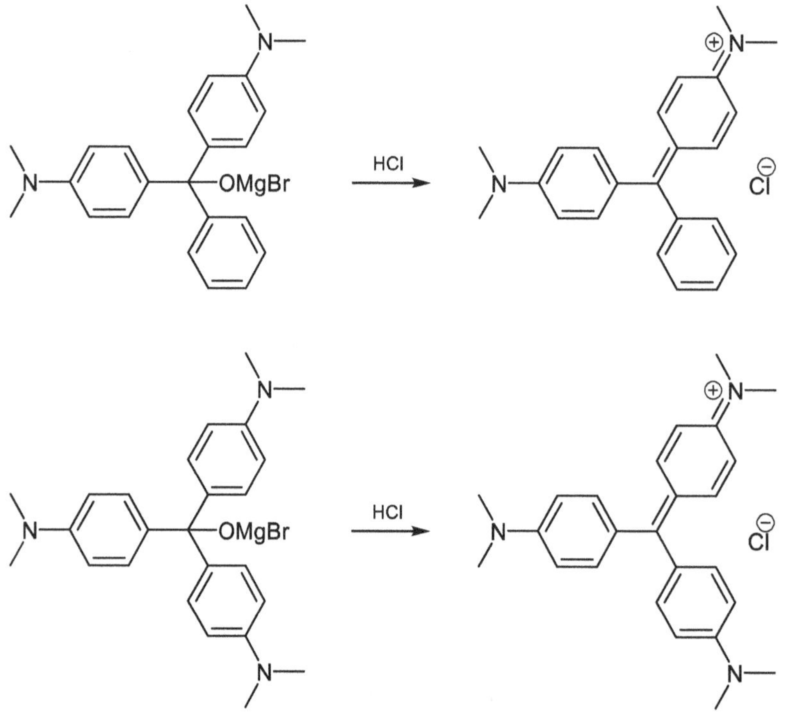

What are the mechanisms of these reactions?

Techniques used, from Zubrick, 8th Ed: two-layer separation (Chapter 15), reflux and addition under water-free conditions (Chapter 22)

Minimum Safety Standards for this experiment

1. Hot glass looks the same as cold glass. Be sure that a piece of glassware is cool enough to handle before picking it up.

2. 4-Bromo-*N,N*-dimethylaniline is an irritant. Wash your hands after contact.

3. Diethyl carbonate is flammable. Use only in the hood.

4. Methyl benzoate is a smelly irritant, and may make you sneeze. Use only in the hood.

5. Malachite green is mildly toxic, and will stain skin and clothing indelibly. Use gloves to handle.

6. Crystal violet is toxic, and will stain skin and clothing indelibly. Use gloves to handle.

7. THF is smelly and flammable. Use only in the hood and wash thoroughly after handling.

8. 10% hydrochloric acid is corrosive. Handle with care and wash thoroughly after handling.

9. Magnesium is flammable and must not be used in the vicinity of open flames.

10. Iodine is a severe irritant. Wash your hands after handling.

11. Drierite is completely innocuous; just keep the dust out of your glassware other than the drying tube.

Disposal

Product solutions	Place in the waste bottle provided.
Diethyl carbonate	Small residues may be flushed down the sink in the hood.
Methyl benzoate	Spills should be wiped with paper towels that are left in the hood until the odor has dissipated. Trace quantities can be flushed down the sink in the hood.
10% hydrochloric acid	Flush down the sink. Clean spills with wet paper towels.
N,N-dimethylaniline	Spills may be swept up and thrown in the trash.
Iodine	Spills may be swept up and thrown in the trash.
Drierite	Place in the trash.

Procedure

Students will be assigned to synthesize either malachite green or crystal violet.

Get a 100-mL flask and a condenser out of the oven. (After lab, you are expected to clean your glassware, rinse it well with water, and return it to the oven.)

After removing your glassware from the oven, you must work quickly to avoid moisture condensing inside your glassware. You should probably weigh out your solid reagents before going to the oven.

1. Place 0.4 g magnesium, reserving one piece and 2.5 g of 4-bromo-*N,N*-dimethylaniline in your 100-mL round-bottomed flask. Add 25 mL of dry THF and a small crystal of iodine to the flask. Crush your reserved magnesium with a stirring rod in a plastic weighing boat and add it to the flask.

2. Fit the flask with a reflux condenser and a drying tube. Reflux the mixture for 30 minutes, being careful not to overheat. (The Grignard reaction takes heating to start, but is exothermic.) Watch for color changes.

3. DO NOT DISASSEMBLE YOUR APPARATUS! Cool the reaction flask to room temperature by using an ice bath. Combine the appropriate amount of your desired ester with about 2 mL of THF in a 25-mL round-bottomed flask **that you get out of the oven**. Add the mixture to the flask through the condenser; put the drying tube back on top of the condenser and reflux for another 5 minutes. Again, cool to room temperature using the ice bath.

   Malachite Green: Use 0.43 g methyl benzoate, measured by volume in the hood using a syringe.

   Crystal Violet: Use 0.25 g diethyl carbonate, measured by volume in the hood using a syringe.

4. Add 8-10 mL of 10% HCl[1] through the condenser, **a little at a time** (the reaction with the excess magnesium is quite exothermic!) The result will be a mixture of the appropriate color, either green or purple.

5. Each student will be given cotton cloth to dye. Immerse the cotton in your dye solution for one minute, then rinse under running water until the water runs clear. Allow to dry on a pad of newspaper or paper toweling.

   You can get interesting patterns by twisting the cloth and using different dyes on different areas. You can also apply the dye with a brush; you might paint a design on your cloth, using one or both colors.

   Caution: The dyes will stain your skin and clothing! Use gloves, and don't wear clothing that you care about! These dyes are not fixed, and so cannot be used on wearable, washable clothing!

For the report

Draw the mechanism of this reaction, beginning with the Grignard reagent and ending with the dye that you synthesized.

What is the atom economy of the synthesis you performed?

Pre-laboratory questions

1. Normally, Grignard reactions give alcohols after aqueous workup of the reaction mixture. Why do we get the products that we do?

2. Most organic compounds are white. Why are malachite green and crystal violet intensely colored? What UV-visible wavelengths do they absorb?

3. Are there any extraordinary hazards present in this experiment? Defend your answer. Consider not only the chemicals used, but also their amounts and the conditions.

---

[1] At this point you will be adding water to the flask, so you don't need the drying tube any more.

Synthesis of an Ether Using Phase-Transfer Catalysis

Ethers may be synthesized by a nucleophilic substitution reaction, in which an organic oxide anion reacts with an alkyl halide to produce a new carbon-oxygen bond. This reaction is called the *Williamson Ether Synthesis*. In this experiment, you will synthesize 1-ethyl-4-propoxy-benzene, an alkyl aryl ether.

The main problem in this synthesis is bringing the phenoxide nucleophile, which is polar, into contact with the 1-iodopropane electrophile, which is non-polar. This difficulty is solved by the use of tetramethylammonium bromide as a *phase-transfer catalyst.*[1]

Techniques used from Zubrick, 9th Ed: microscale reflux (Chapters 22-23), microscale extraction and washing (Chapter 16). Drying in a drying column and evaporation as we are doing it are not discussed in Zubrick.

Minimum Safety Standards for this experiment

1. Hot glass looks the same as cold glass! Before picking up a piece of glassware, be sure to check that it is cool enough to handle.

2. 4-Ethylphenol is a toxic, smelly irritant (it may make you sneeze). Spills should be cleaned with soap and water *in the hood*. This reagent must be kept inside the hood at all times! Weighing *must* be performed in closed containers. Nothing contaminated with this reagent may be removed from the hood until the odor is gone. The smell will probably permeate your clothes and skin by the end of the experiment.

3. 1-Iodopropane is toxic and carcinogenic. The quantities we will use are relatively safe. Keep this reagent in the hood.

4. Tetramethylammonium bromide is toxic and should be treated with respect. Because it will absorb water from the air, the container should be kept tightly closed when not in use. This reagent may be used outside the hood.

5. You will be working with concentrated solution of sodium hydroxide. This solution is quite caustic and should be handled with great care! Wash thoroughly after handling.

6. Diethyl ether has no major hazards associated with it **except extreme flammability**. Keep away from flames and use **only** in the hood; excessive exposure can cause narcosis.

7. Magnesium sulfate is sold as "Epsom Salts." Alumina is non-toxic (inhalation of large amounts leads to mechanical lung damage, or *silicosis*). There are no significant hazards associated with these substances in the quantities used for this experiment.

---

[1] "Phase-transfer catalysis" is discussed in Loudon; see the index.

Disposal

Alumina	Spillage may be swept up and placed in the wastebasket.
Diethyl ether	Most of the ether will be evaporated. Wipe up spillage with paper towels and allow the ether to evaporate in the hood. The paper towels may then be disposed of in the wastepaper basket.
Drying column	May be disposed of in the container for used Pasteur pipettes.
4-Ethylphenol	Residues on glassware should be cleaned with acetone and flushed down the sink **in the hood.** Spills may be cleaned with soap and water, with the wash water kept **in the hood.**
1-Iodopropane	Residues should be cleaned with acetone and flushed down the sink.
Magnesium sulfate	Spillage may be swept up and placed in the wastebasket.
4-propoxyethylbenzene	Product residues should be cleaned with acetone and flushed down the sink.
Sodium hydroxide solutions	Clean with water and flush down the sink.
Tetramethylammonium bromide	Residues should be cleaned with soap and water and flushed down the sink.

Procedure

1. Weigh 100-200 mg 4-ethylphenol into a 3-mL conical vial with a spin vane. Add about 30 mg tetramethylammonium bromide, 0.25 mL 25% aqueous NaOH and 1 µL 1-iodopropane per mg of 4-ethylphenol.[1] Attach a condenser.

2. Heat the reaction mixture on medium at 80-95°C for 60 minutes with vigorous stirring. Measure the temperature by *carefully* inserting a thermometer bulb into the small hole in the heating block.

   DO NOT ALLOW YOUR MIXTURE TO OVERHEAT. What is the boiling point of 1-iodopropane? Keep your block temperature under that value or you will lose it all!

3. Allow the reaction to cool and remove the spin vane. Two phases should be apparent: an upper organic layer and a lower aqueous layer. Add 0.5 mL diethyl ether, shake, and allow the layers to separate again.

4. Separate the layers. Extract the aqueous layer with another 0.5 mL diethyl ether. Discard the aqueous layer, and combine the ether layers.

5. Wash the organic layer with about 0.25 mL 5% aqueous NaOH, then with about 0.25 mL water.

---

[1] That is, if you have 125 mg of 4-ethylphenol, you need 125 µl = 0.125 ml of 1-iodopropane..

114

6. Add 0.25 mL ether to the organic layer and pull the ether solution through a drying column containing magnesium sulfate (1 cm) over alumina (2 cm) in a Pasteur pipette, collecting the material in a tared conical vial. Wash the column with 3 mL of ether, pulled through into the tared vial.

7. Evaporate the solvent using a clean Pasteur pipet, vacuum adapter and aspirator, set up as before (minus the drying agent). Determine the mass of your product and report the yield.

8. Analyze your product by ^{1}H NMR and COSY.[1] Prove the structure of your product from its spectra.

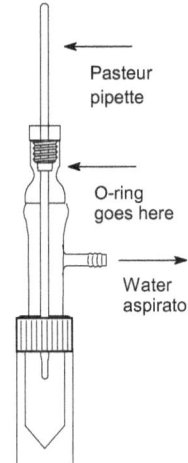

For the report
Prove the structure of your product from the NMR spectra. Draw a detailed reaction mechanism, and discuss the role of the phase-transfer catalyst in this reaction.

Pre-laboratory assignment
4. Draw the structures of your starting materials and the expected product.

5. What features do you expect to see in the hydrogen NMR spectrum of the product?

6. Are there any extraordinary hazards present in this experiment? Defend your answer. Consider not only the chemicals used, but also their amounts and the conditions.

Oxidation of a bifunctional alcohol

Traditionally, chromium(VI) (Cr^{6+}) compounds have been used for the oxidation of alcohols. However, chromium(VI) is toxic and both use and disposal of its compounds are hazardous. Therefore, procedures have been recently developed which use more environmentally benign oxidizing agents. In this experiment we will use sodium hypochlorite (NaOCl), the active ingredient in household bleach, as the oxidizing agent. Hypochlorite (or in this experiment, hypochlorous acid) functions as a source of "Cl^+," converting alcohols into organic hypochlorites that can then undergo elimination to give carbonyl groups:

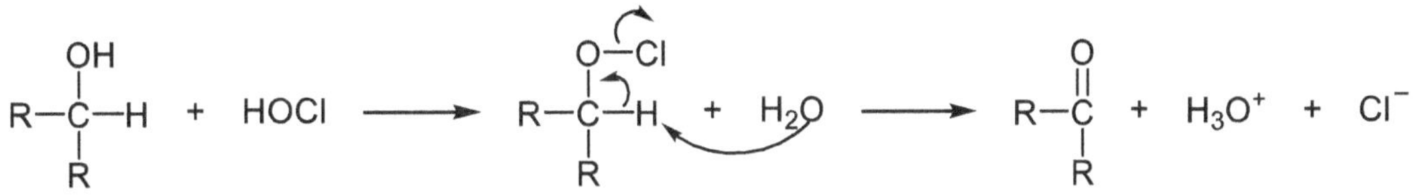

One purpose of this experiment is to determine the specificity of hypochlorite/hypochlorous acid as an oxidizing agent for alcohols. In order to do this we will oxidize the bifunctional compound 2,2,4-trimethylpentane-1,3-diol.[2] There are three possibilities: both alcohols could be oxidized; only the secondary alcohol could be oxidized; only the primary alcohol could be oxidized. The primary alcohol, if oxidized, could be oxidized to either an aldehyde or to a carboxylic acid.

[1] See the section, "Introduction to two-dimensional NMR spectroscopy."

[2] This experiment adapted from Pelter, M.W.; Macudzinski, R.M.; Passarelli, M.E. *J. Chem. Educ.* **2000**, *77*, 1481.

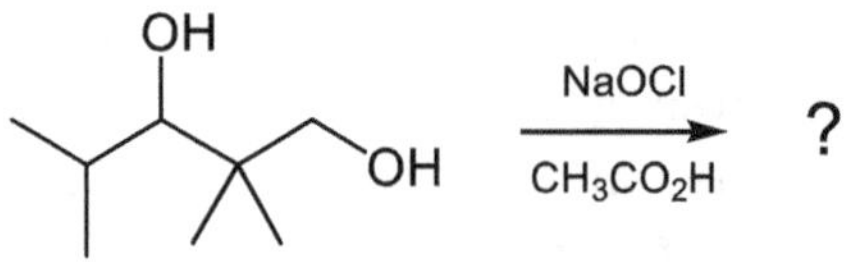

OH ... OH, NaOCl / CH$_3$CO$_2$H → ?

This experiment is time-consuming because of the extended reaction time. Be sure to work efficiently and plan thoroughly, or you may have to leave the laboratory without finishing. This will result in a grade of zero!

Techniques used from Zubrick, 9th Ed: extraction and washing (Chapter 15), drying an organic liquid (Chapter 10)

Minimum Safety Standards for this experiment

1. Hot glass looks the same as cold glass! Before picking up a piece of glassware, be sure to check that it is cool enough to handle.

2. Reagents which have an odor or an appreciable vapor pressure may not be used outside the hood except in closed containers.

3. Look up the MSDS for each reagent used. More specific cautions and procedures are given below.

4. Household bleach is caustic and a strong oxidizer; glacial acetic acid is corrosive. Wash thoroughly after handling.

Disposal

drying agent	Solid spillage may be swept up and placed in the wastebasket. Used drying agent must be kept in the hood until the odor is gone, then placed in the wastebasket.
ether	Most of the ether will be evaporated. Wipe up spillage with paper towels and allow the ether to evaporate in the hood. The paper towels may then be disposed of in the wastepaper basket.
glacial acetic acid	Residues should be cleaned with water and flushed down the sink **in the hood**.
household bleach	Clean with soap and water and flush down the sink.
reaction product	Residues should be cleaned with acetone and flushed down the sink. Leftover product, if any, should be placed in the bottle for non-halogenated organic waste.
sodium carbonate solution	Flush down the sink.
sodium chloride solution	Flush down the sink.
sodium hydroxide solution	Flush down the sink.
2,2,4-trimethylpentane-1,3-diol	Residues on glassware should be cleaned with acetone and flushed down the sink. Spills should be swept into the trash, then cleaned with soap and water.

Procedure

1. Place 0.75-1 g of the alcohol into a 50-mL round-bottomed flask. Determine the mass of the alcohol carefully. Add about 1 mL of glacial acetic acid and insert a magnetic stirring bar.

    Glacial acetic acid may be used ONLY in the hood!

2. Stir the solution using a magnetic stirrer. Add about 8 mL of 5% sodium hypochlorite (NaOCl) solution (household laundry bleach!) a little at a time, over a 2-5 minute period.

3. Stopper the flask and stir vigorously at room temperature for 60 minutes. Periodically check your solution with starch-iodide paper[1] to ensure that it contains excess hypochlorite; if the starch-iodide test is negative (no color change), add more bleach, about 0.5 mL at a time, until a positive test is obtained.

4. After the hour is up, pour your mixture into a separatory funnel with about 10 mL of brine;[2] you may want to remove the stir bar first! Extract three times with 5-7 mL portions of diethyl ether.

5. Wash the combined extracts with three 3-to-5 mL portions of 5% sodium carbonate followed by two 3-to-5 mL portions of 5% aqueous NaOH. Wash the ether layer with about 1 or 2 mL of water; if the water layer is acidic to litmus paper, you will need to perform more washings with 5% NaOH. You should add more ether if your organic layer goes below about 10 mL in volume.

6. Dry the combined ether extracts over calcium chloride. Transfer the ether layer to a clean, water-free round-bottomed flask, and remove the ether on the rotary evaporator.

7. Determine the yield of your product. Analyze your product by IR and NMR. What are the possible products? Which of them have you made?

For your report

Calculate the yield of this and every synthesis you perform. Discuss the spectroscopic results, and show which of the possible products they indicate. Might it be possible to determine the structure of the product by IR alone?

Why did you wash the reaction mixture with strong base? Consider both reaction mixture and possible products!

Pre-laboratory assignment

1. Draw **all** of the possible oxidation products. Consider: (a) primary alcohols may be oxidized to aldehydes or to carboxylic acids; (b) both alcohols may be oxidized, or only one.

2. Explain how you would distinguish between the different possible oxidation products by hydrogen NMR, and by IR spectroscopy.

3. Are there any extraordinary hazards present in this experiment? Defend your answer. Consider not only the chemicals used, but also their amounts and the conditions.

---

[1] Dip a stirring rod or spatula, NOT the test paper!

[2] What is "brine"?

The Aldol Condensation

In the Aldol condensation, strong base is used to convert an aldehyde or ketone (the Aldol *donor*) into an enolate, which then attacks the carbonyl group of an aldehyde (the Aldol *acceptor*). Finally, the adduct is dehydrated to produce a double bond. The reaction is often performed with the same compound acting as donor and acceptor, but not always.

In this experiment, we will use acetone as the donor and benzaldehyde as the acceptor. Notice that acetone has two identical α-carbons, each of which can lose a proton to create an enolate nucleophile. Therefore, it is possible for each acetone molecule to react with *two* benzaldehyde molecules. The actual product of this reaction, either 4-phenyl-3-buten-2-one or 1,5-diphenyl-1,4-pentadien-3-one, depends on the reaction conditions. You will be expected to determine the identity of the product obtained.

Techniques used from Zubrick, 9th Ed: vacuum filtration and recrystallization (Chapter 13)

Minimum Safety Standards for this experiment

1. 1.5 M NaOH is caustic. Use care in handling and wash thoroughly afterward.

2. Benzaldehyde is a smelly irritant. Use only in the hood, and wash thoroughly after handling.

3. Acetone and ethanol are not toxic externally.

Disposal

All materials used in this experiment may be flushed down the sink. Large spills of benzaldehyde must be wiped up with a paper towel, and the paper towel allowed to dry in the hood before disposal.

Leftover product may be discarded in the wastebasket.

Procedure

1. To a 25-mL Erlenmeyer flask, add about 6 mL of sodium hydroxide solution[1] and a magnetic stirring bar. Now add 0.37 mL acetone, measured precisely with an autopipette.

2. With moderate stirring, add 1.00 mL benzaldehyde – again, measured with an autopipette. Continue to stir the mixture for 30-45 minutes[2] at room temperature. Record your observations.

3. Collect the product by vacuum filtration. Wash the product with water, alternating with *small* amounts of 60/40 ethanol/water. Be careful not to lose the product!

4. After washing, pull air through the product for 10 minutes to dry it. If there is time, you may recrystallize from 95% ethanol. Allow your product to stand for at least 24 hours before weighing it, to allow it to finish drying.

---

[1] The solution is 1.5 M NaOH in 40/60 ethanol/water.

[2] Use your judgement, depending on what you observe!

118

5. Obtain the mass, melting point, and IR and NMR spectra of your product. Identify it and determine the crude and final yields. How might you change the reaction conditions in order to increase the yield of the other possible product?

Pre-laboratory assignment

1. Explain how you will distinguish the two possible products by NMR.

2. Draw a complete reaction mechanism for the formation of either one of the two possible products.

3. Are there any extraordinary hazards present in this experiment? Defend your answer. Consider not only the chemicals used, but also their amounts and the conditions.

For the report

Explain the mechanism of the reaction and how you distinguished the two possible products from each other.

Dieckmann Condensation of Adipic Acid

The Dieckmann condensation is an enolate addition reaction in which one end of a molecule acts as nucleophile toward the other end, forming a ring. For example, dimethyl heptanedioate can be reacted with itself to form methyl 2-oxocyclohexanecarboxylate:

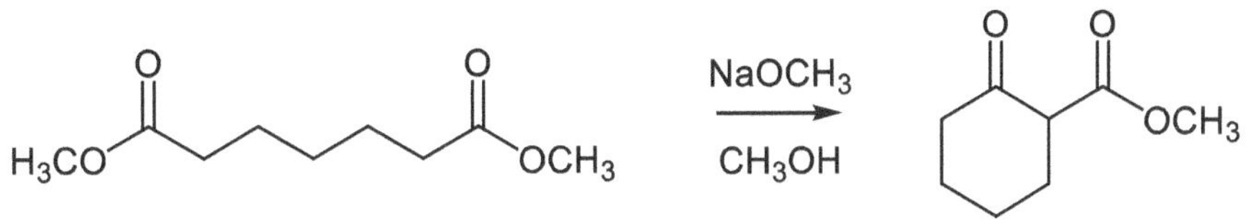

In situations in which two possible ring sizes may form, we find that 5- and 6-membered rings are favored over other ring sizes.

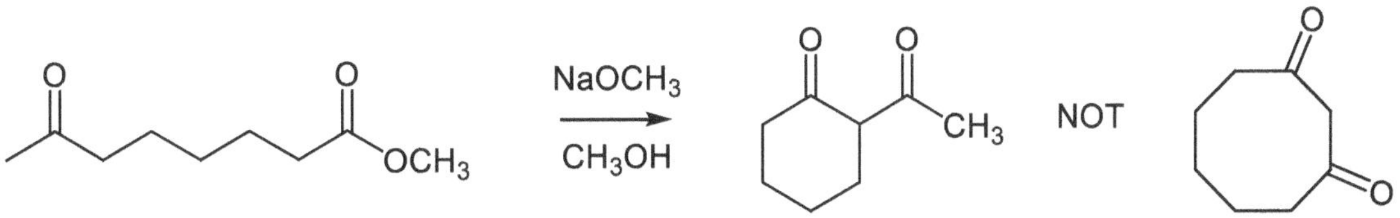

Which open-chain enolate reaction is the Dieckmann condensation similar to: the aldol condensation or the Claisen condensation?

Minimum Safety Standards for this experiment

1. Hot glass looks the same as cold glass! Before picking up a piece of glassware, be sure to check that it is cool enough to handle.

2. Barium hydroxide is toxic; the lethal dose for an adult is estimated at between 1 and 15 grams. The good news is that it must be ingested to be toxic; however, like any strong base, it is a severe irritant when powdered. You must grind your barium hydroxide in the hood to protect yourself from inhaling the dust!

3. Sodium hydroxide in alcohol is highly caustic and will eat holes in both skin and clothing. Wash your hands immediately after handling.

4. Calcium chloride is an aggressive dessicant; avoid contact. Wash your hands after handling.

5. Adipic acid is not poisonous but, like any acid, is a severe eye irritant. Avoid contact with eyes.

6. Cyclopentanone is a skin and eye irritant; avoid contact. Quantities greater than 1 mL must be kept in the hood or in a closed container.

Disposal

Adipic acid	Residues should be cleaned with acetone or soap and water, and flushed down the sink.
Aqueous layer from extraction	Clean with water and flush down the sink.
Barium hydroxide residues	Clean with water and flush down the sink.
Calcium chloride	Spills may be swept up and placed in the trash. The area of the spill should then be wiped with wet paper towels, which may also be placed in the trash.
Cyclopentanone	Residues should be cleaned with acetone or soap and water, and flushed down the sink.
Distillation residue	The black polyester residue from distillation should be soaked for at least 30 minutes in 6M NaOH, then removed, with scraping. You may need to soak the flask overnight. After getting them out of the flask, soak the solids in one or two changes of water, then discard in the trash.
Sodium hydroxide, saturated in ethanol	*Small* residues (less than 100 mL) may be cleaned with water, and flushed down the sink. Larger amounts must be neutralized before disposal.

Procedure

1. Using a dry mortar and pestle, grind together 10-15 grams of adipic acid with about $^1/_5$ its weight of barium hydroxide until the mixture is a fine powder. Transfer the powder to a round-bottomed flask and add two or three boiling chips.

2. Assemble a simple distillation apparatus, using the flask with the barium hydroxide mixture as the still pot and placing the receiving flask in an ice/water bath. Use a *light* coat of petroleum jelly to

prevent the ground glass joints from freezing.[1] Heat the flask gently until the powder is completely melted, then heat more strongly until liquid begins to distill. Continue to distill slowly, keeping the still head temperature below 150°, until only solid residue is left in the flask.

After the distillation is complete: To keep the standard-taper joints from freezing, loosen them carefully as soon as possible, using heat-protective gloves. However, **do not** disassemble the apparatus until it has completely cooled. Protect the thermometer from sudden temperature changes that might break it by leaving it in place until the apparatus has cooled.

Break up the residue in the boiling flask and wash out the apparatus **as soon as possible after use**, being especially careful to clean the joints thoroughly. Soak the still pot for 30 minutes with ethanolic NaOH solution, then scrape and rinse thoroughly.

3. Transfer the distillate to a separatory funnel, washing the receiving flask with 3 mL of saturated aqueous sodium chloride. Shake the mixture to "salt out" the organic layer and allow it to stand. When the layers are well separated, remove the aqueous layer completely and dry the organic layer over anhydrous calcium chloride.

4. Purify your product by simple distillation, using standard-scale distillation or a microscale (Hickman-Hinkle) distillation depending on the quantity of your product (less than 5 mL requires a microscale distillation). Distill over a range of about 128-135°.

5. Obtain the yield and spectra of your product. The expected product of this reaction is cyclopentanone.

For the report...

Report the yield. Justify your identification of the product as cyclopentanone. Suggest a plausible mechanism for the conversion of adipic acid into cyclopentanone under these conditions.

Pre-laboratory assignment:

1. Propose a detailed mechanism for the Dieckmann condensation reaction of dimethyl heptanedioate, shown in the introduction to this experiment. Please note that this is not the particular Dieckmann condensation you will carry out!

2. Be sure to read about microscale distillation techniques in Zubrick.

3. Barium hydroxide is toxic when taken by mouth, but barium sulfate is commonly used as an oral x-ray contrast agent for digestive system diagnosis. Why is barium hydroxide toxic, while barium sulfate can be eaten by the pound? (NOTE: it has nothing to do with hydroxide vs. sulfate; both anions are non-toxic. It is barium that is toxic.)

4. Are there any extraordinary hazards present in this experiment? Defend your answer. Consider not only the chemicals used, but also their amounts and the conditions.

---

[1] You do **not** need to lubricate the joints connecting the condenser to the receiving flask, and in fact such lubrication may lead to contamination of your product.

A Synthesis Using Meldrum's Acid[1]

In this experiment, you will identify the product of a reaction between Meldrum's acid and formaldehyde.

Techniques used from Zubrick, 9th Ed: microscale glassware (Chapter 5), vacuum filtration and recrystallization (Chapter 13), NMR (Chapter 33)

Minimum Safety Standards for this experiment

1. Meldrum's acid is an irritant. Treat it with respect.

2. Dimethylformamide (DMF) is smelly, an irritant and a teratogen. It will penetrate latex gloves and go through human skin, along with anything dissolved in it. Avoid direct contact and wash your hands thoroughly after using.

3. 37% formaldehyde is smelly, a lachrymator and possible carcinogen; however, the amounts used in this experiment are small and you will not handle the concentrated reagent directly.

Disposal

1. Spilled Meldrum's acid may be discarded in the trash.

2. The aqueous waste produced in this reaction, as well as DMF and formaldehyde rinsings, must be flushed down the sink *in the hood*.

3. Large DMF spills must be cleaned up with paper towels. These towels must remain in the hood until their odor is gone. Wash thoroughly after cleaning up such a spill.

4. Chloroform-*d* solutions must be placed in the appropriate waste bottle.

5. Leftover product may be discarded in the trash.

Procedure

1. Place 1.00 mmol[2] of Meldrum's acid (2,2-dimethyl-1,3-dioxane-4,6-dione) in a 3-mL conical vial and add about 200 μL of DMF. Add a spin vane and stir gently to dissolve the solid.

2. Add 38 μL of 37% aqueous formaldehyde.[3] Cap the vial and stir for 90 min at room temperature.

3. Add about 500 μL of tap water and cool in an ice bath for 10-15 min.

4. Collect the product by filtration, using a Hirsch funnel. Rinse the vial with a *minimum* amount of *cold* water. Solid will form in the filter flask and a second crop of product can be obtained from the filtrate to increase your yield.

---

[1] Based on Crouch, R.D.; Holden, M.S. *J. Chem. Educ.* **2002**, *79*, 477-478.

[2] It is not necessary to get *precisely* 1.00 mmol, but you should (a) get as close as possible and (b) record the *exact* amount of Meldrum's acid you use.

[3] This will be dispensed by the instructor. Adjust the amount of formaldehyde based on the actual amount of Meldrum's acid used.

5. Dry your product in the drying cabinet. Record the mass and melting point of your dried product and obtain appropriate NMR spectra.[1]

For the report

Identify your product; it forms according to the stoichiometry in the reaction conditions. Discuss its spectral data (especially NMR). Propose (and draw) a reasonable mechanism for the formation of your product.

Pre-laboratory assignment

Why do they call it "Meldrum's acid"?

You know, of course, the starting materials and conditions. What products are possible? Likely? Consider the stoichiometry of the reaction![2]

For guidance, see the sections of your text on carbonyl alpha-substitution reactions.

Synthesis of Bisphenol Z

This experiment is taken from Gregor, R.W. *J. Chem. Educ.* **2012**, *89*, 669-71.

Polycarbonate is one of the most durable and useful artificial polymers. It is typically synthesized from the reaction of phosgene and the dianion of bisphenol A, which has the IUPAC name 4,4'-(propane-2,2-diyl)diphenol:

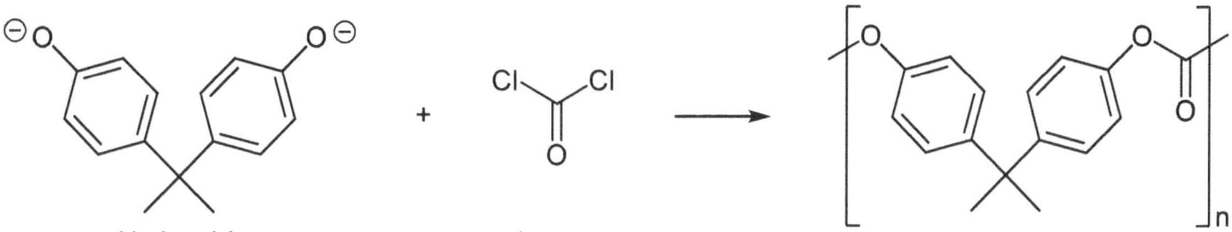

bisphenol A
dianion

phosgene

Bisphenol A, in turn, is easily made from acetone and phenol in the presence of acid, a reaction known as electrophilic acyl addition. You should try to work out a mechanism for this. HINT: draw resonance structures for phenol, and consider that, in the presence of acid, a carbonyl group is strongly activated toward reaction with nucleophiles.

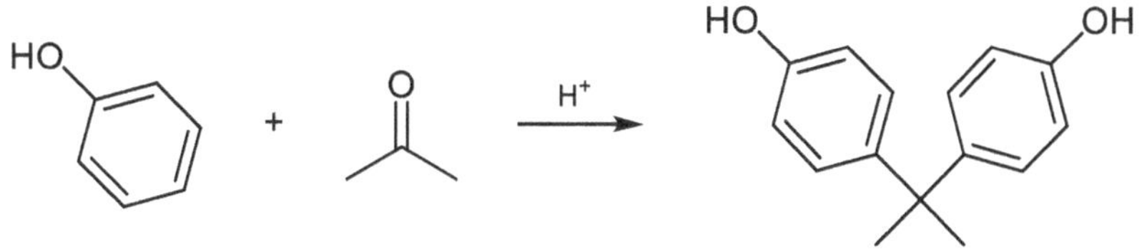

---

[1] Literature values: m.p. 144-146°C (Hedge, J. A.; Kruse, C. W.; Snyder, H. R. *J. Org. Chem.* **1961**, *26*, 3166); ^{1}H NMR: [CDCl$_3$] δ 4.53 [t, J = 1.76 Hz, 2H], 2.79 [t, J = 1.76 Hz, 2H], 1.84 [s, 6H], 1.79 [s, 6H]; ^{13}C NMR: [CDCl$_3$] δ 165.4, 105.7, 42.5, 24.5, 26.4, 23.2 (Crouch, R.D.; Holden, M.S. *J. Chem. Ed.* **2002**, *79*, 477).

[2] 37% formaldehyde is a solution in which each milliliter contains 0.37 grams of formaldehyde (37 grams per 100 mL).

Polycarbonate is a glassy, durable hard thermoplastic that is used for objects from aircraft windows to water bottles. However, when overheated in the presence of water, polycarbonate can <u>slowly</u> break down to carbonic acid and bisphenol A, which can in turn leach out of the material. Bisphenol A is a known – though extremely weak[1] – endocrine disruptor; it is similar in structure to female hormones such as estradiol. However, polycarbonate is *such* a useful material that there is great reluctance – on the part of everybody – to do without it, endocrine disruptor or not.

The reaction we will perform will be the synthesis of bisphenol Z, 4,4'-(cyclohexane-1,1-diyl)diphenol, from phenol and cyclohexanone. This reaction is exactly similar to the synthesis of bisphenol A, with cyclohexanone replacing phenol. The reason for using cyclohexanone instead of acetone is that cyclohexanone is far less volatile, and so less likely to escape during the reaction.

The bisphenol family of reactions tend to form 1:1 adducts between the bisphenol product and phenol, which is present in excess. These adducts may be isolated and characterized. The components are likely held together by hydrogen bonds. You will explore the nature and strength of this interaction through thin layer chromatography (TLC), melting point (mp), and ^{1}H NMR. You will then separate, recrystallize, and characterize the pure bisphenol Z product.[2]

Minimum Safety Standards for this experiment

1. Hot glass looks the same as cold glass. Be cautious when handling glassware.

2. Concentrated HCl (37% by weight) is corrosive and emits corrosive, irritating fumes. Use it in the hood, and clean up spills promptly with water.

3. Phenol is mildly corrosive and will burn the skin and eyes. Use it with caution; wear gloves, or wash hands thoroughly after handling. Clean up spills promptly.

4. Toluene is inflammable, teratogenic and mildly intoxicating when inhaled. Use only in the hood, and inform the instructor if you are pregnant or trying to become pregnant.

5. Cyclohexanone is inflammable, and should be used only in the hood.

6. Dichloromethane is volatile and mildly carcinogenic. Use only in the hood.

---

[1] See Sharpe, R.M. "Is It Time to End Concerns over the Estrogenic Effects of Bisphenol A?" *Toxicol. Sci.* **2010**, *114*, 1-4, doi: 10.1093/toxsci/kfp299. This editorial from *Toxicological Science* cites several studies showing weak or absent estrogenic effect from large oral doses of bisphenol A.

[2] Bisphenol Z is in the SDBS under "4,4'-cyclohexylidenediphenol." Its SDBS number is 19262. Its melting point may be found in the Aldrich Catalog under "4,4'-cyclohexylidenebisphenol."

7. Bisphenol Z is a mild endocrine disruptor. Wash hands after handling, and avoid ingestion.

8. Dimethyl sulfoxide-d_6 (DMSO-d_6) is readily absorbed through the skin. Avoid contact with DMSO solutions.

Disposal

bisphenol Z	Residues may be cleaned with acetone or soap-and-water, and flushed down the sink. Your product should be placed in the waste bottle provided.
concentrated HCl	May be flushed down the sink.
cyclohexanone	Spills may be wiped up with paper towels and allowed to evaporate in the hood. Place used cyclohexanone in the waste bottle provided. Flush residues down the sink.
dichloromethane	Spills may be wiped up with paper towels and allowed to evaporate in the hood. Place used dichloromethane in the waste bottle provided. Flush residues down the sink.
DMSO-d_6	Place used NMR solutions in the NMR waste bottle. Flush residues down the sink.
filtrate (liquid from filtration)	Any mixture containing significant amounts of toluene must be placed in the appropriate waste bottle. If the mixture is more than 90% aqueous, you may flush it down the sink *in the hood*. These liquids will contain significant amounts of phenol, so be sure to *flush* the liquid down the sink.
methanol	Methanol solutions may be flushed down the sink.
phenol	Spills should be swept up and placed in the trash. For spills larger than a few hundred milligrams, ask the instructor for advice.
toluene	Spills may be wiped up with paper towels and allowed to evaporate in the hood. Place used toluene in the waste bottle provided. Flush residues down the sink.

Procedure

Prepare a 70°C warm water bath using a 250 mL beaker, about half full. Place 12 mmol phenol, 3 mmol cyclohexanone, and 0.5 mL of concentrated HCl in a 25 mL round-bottomed flask with a small magnetic stirring bar. You should observe two distinct layers: a cloudy white bottom layer of about 0.5 mL and a pale yellow top layer of about 1.5 mL. Attach a water-cooled reflux condenser and place the assembly in the hot water bath. Magnetically stir about 1 hour at 70°C. The color should change from a pale yellow to a dark orange or even brown. Solid product should be plainly visible, especially after cooling.

Cool the reaction mixture in an ice bath and filter the contents of the flask using a Buchner funnel attached to an aspirator through a 125 mL side-armed flask. Using a small spatula to help break up the solid product and to facilitate drying. Wash on the filter paper with about 30 mL of water followed with about 10 mL toluene,[1,2] until the color is white to light pink. (Use more toluene if the product has an

---

[1] Remember that using your wash in several small portions will do a much better job than using it all in one slug!

[2] The liquid filtered away from your solid must be separated before disposal; put the organic layer in the appropriate waste bottle.

orange tint.) After drying on the aspirator, weigh the bisphenol Z/phenol adduct, and reserve samples for melting point, TLC, and ^{1}H NMR.

Thoroughly mix the adduct with about 30 mL of boiling water in a beaker, and then filter as before to remove the phenol. Wash the solid with several small portions of hot water. After drying on the aspirator, weigh the crude bisphenol Z product.

Recrystallize your bisphenol Z by dissolving it in a <u>minimum</u> of hot methanol (a few mL or less), cooling in an ice bath, and then filtering using the same filtration setup. After washing the solid with a <u>small</u> amount of ice cold methanol, dry the product on the filter. Weigh the pure bisphenol Z product and reserve samples for melting point, TLC, and 1H NMR.

Run a TLC plate spotted[1] with phenol, the adduct, and the final product using dichloromethane-ethyl acetate (90:10) as the eluent. Obtain ^{1}H NMR spectra of the adduct and of the final product, using DMSO-d_6 or acetone-d_6 as the NMR solvent.

Pre-laboratory questions

1. Calculate masses required for all reagents. Calculate the volume of cyclohexanone required; you can find the density with Google.

2. Find the ^{1}H NMR spectra for phenol and cyclohexanone.[2] Find the ^{1}H NMR chemical shifts for the following possible impurities: DMSO (dimethyl sulfoxide), methanol, H_2O, and toluene. If possible, find these numbers for spectra in DMSO-d_6.

3. Propose a mechanism for the synthesis of bisphenol A or Z from phenol and acetone or cyclohexanone under acidic conditions. HINT: phenol has an important resonance structure that has positive charge on oxygen and negative charge on the #4 carbon of the ring.

4. What are the industrial sources of phenol, acetone, and cyclohexanone? How do they rank in worldwide (or U.S.) chemical production?

For the report

1. Report R_f values for phenol and bisphenol Z. When determining the yield of your bisphenol Z/phenol adduct, assume that the mole ratio is 1:1. Compare the ^{1}H NMR spectra of phenol,[133] the adduct and the purified product, and discuss any differences in chemical shifts caused by interactions in the adduct.

2. Estimate the actual bisphenol Z to phenol mole ratio in your adduct based on the masses of crude adduct and crude bisphenol Z.

3. Discuss the separation of bisphenol Z from phenol in both the bulk and on the TLC plate in terms of solubility and binding.

---

[1] All the solids on which you are to run TLC will dissolve in acetone for spotting.

[2] The spectra of cyclohexanone, phenol and bisphenol Z (4,4'-cyclohexylidenediphenol) are available on SDBS.

The Suzuki Coupling Reaction

Based on E. Aktoudianakis et al, *J. Chem. Educ.* **2008**, *85*, 555-557

Palladium-catalyzed cross-coupling reactions won the Nobel Prize in Chemistry in 2010 and are a staple of synthetic organic chemistry. They have become standard ways of making a variety of carbon-carbon bonds, and are used in the synthesis of a number of different compounds, including a number of pharmaceuticals such as ibuprofen.

Biaryl compounds, which can be made via the Suzuki coupling, are common among non-steroidal anti-inflammatory drugs. Two examples that are similar to the product we are making are felbinac and diflunisal, both of which are useful treatments for arthritis.

felbinac   diflunisal

In this experiment, we will couple phenylboronic acid with 4-iodophenol to make 4-phenylphenol. The reactants and product have been chosen for their low toxicity and ease of identification.

Minimum safety standards for this experiment

1. Hot glass (and other equipment) looks the same as it does when it is cold. Since you are running the reaction at or above the boiling point of water, you must take care not to burn yourself.

2. Phenylboronic acid (benzeneboronic acid) is an irritant. Treat it with respect.

3. *p*-Iodophenol is an irritant. Treat it with respect.

4. Potassium carbonate is mildly caustic but non-toxic. Wash your hands after handling.

5. Palladium on carbon is innocuous, but because it is a fine powder it can cause respiratory difficulties if inhaled.

6. 2M Hydrochloric acid is corrosive but non-toxic. Wash your hands after handling.

7. Methanol is mildly toxic. Avoid skin contact where possible, and wash your hands after handling.

8. *p*-Phenylphenol is an irritant. Treat it with respect.

Disposal

benzeneboronic acid	Spillage should be swept up and placed in the trash. Solutions may be flushed down the sink, cleaning with acetone or isopropanol.
p-iodophenol	Spillage should be swept up and placed in the trash. Solutions may be flushed down the sink, cleaning with acetone or isopropanol.
potassium carbonate	Spillage should be swept up and placed in the trash. Solutions may be flushed down the sink.
palladium on carbon	Residues on filter paper should be placed in the wastebasket. Residues on glass may be flushed down the sink.
2M hydrochloric acid	Flush down the sink. Clean spills with paper towels, then place in the trash.
methanol	Flush down the sink. Clean spills with paper towels, then place in the trash.
aqueous waste from the reaction	Flush down the sink.
mother liquor from crystallization	Flush down the sink. Clean residues with acetone or isopropanol.
p-phenylphenol product	Place in the wastebasket. Clean residues with acetone or isopropanol.

Procedure

You will be given a vial containing 10 mg of 5% palladium on carbon powder. After use, dispose of the vial in the trash.

This reaction runs at high temperatures, so take care when handling hot equipment!

1. In a 50-ml round-bottomed flask, combine the following: phenylboronic acid (122 mg); potassium carbonate (415 mg); 4-iodophenol (220 mg); and deionized water (10 ml). Add a magnetic stir bar.

2. Add about 1 ml of DI water to the vial containing palladium on carbon powder to create a suspension. Add the suspension to the reaction mixture.

3. Attach a reflux condenser to the flask and heat **vigorously** at reflux for 30 minutes, maintaining rapid stirring. Afterward, remove the flask from the heat and allow it to cool to room temperature.

4. Add about 2 ml of 2M HCl to the mixture. If the mixture is not acidic to litmus paper, add more acid until the mixture tests acidic. Collect the crude solid, which contains the catalyst, by vacuum filtration and wash with 10 ml of DI water.

5. Place the filtered solids in a 25-ml Erlenmeyer flask and dissolve with 10 ml of methanol. Use gravity filtration to remove the palladium/carbon powder, collecting the filtrate in a 50-ml Erlenmeyer flask. If your filtered solution is still black, filter it again; add a little more methanol if needed.

6. Add 10 ml of DI water to the methanol solution, and heat until all solids have dissolved. If necessary, add 1-2 ml of additional methanol.

7. Allow the solution to come to room temperature and crystallize, finishing it in an ice-water bath.

8. Collect the crystalline product by vacuum filtration and allow it to dry for several days. Analyze your product for yield, and use melting point, NMR and IR to characterize it. Compare with literature spectra where possible; spectra of 4-phenylphenol can be found on SDBS.

   NOTE: 4-phenylphenol is not soluble in chloroform; you should dissolve it in acetone-d_6 for NMR. Remember that acetone-d_6 has a solvent peak around 2 ppm, which will be prominent if the solvent has not been freshly opened. Our acetone-d_6 contains 1% TMS, so you do not need to add TMS to your NMR sample.

For the report

Draw a detailed mechanism for the Suzuki coupling reaction that we are using. See Karty, Chapter 19, for a discussion.

Palladium-catalyzed cross-coupling reactions are typically considered "green." What is the atom economy for the coupling we performed? Does it seem particularly green? Why else would this reaction be considered "green"?

Pre-laboratory assignment

1. Read about the Suzuki coupling reaction, and suggest precursors that could be coupled to form felbinac and diflunisal.

2. Calculate the atom economy of this reaction. Is potassium carbonate a stoichiometric or catalytic reactant?

3. During lab, we will discuss catalytic cycles and how they are commonly drawn.

4. Are there any extraordinary hazards present in this experiment? Defend your answer. Consider not only the chemicals used, but also their amounts and the conditions.

The Heck Coupling Reaction

Based on L.L.W. Chung et al, *Chem. Educator* **2007**, *12*, 77-79

Palladium-catalyzed cross-coupling reactions won the Nobel Prize in Chemistry in 2010 and are a staple of synthetic organic chemistry. They have become standard ways of making a variety of carbon-carbon bonds, and are used in the synthesis of a number of different compounds, including a number of pharmaceuticals such as ibuprofen.

The Heck reaction couples an electron-deficient alkene to an aryl halide in the presence of palladium. The general reaction is

What is the purpose of the base?

For example, the common sunscreen 2-ethylhexyl 4-methoxycinnamate is synthesized using a Heck reaction.

In this experiment, we will couple 4-iodoacetophenone with acrylic acid to make *(E)*-4-acetylcinnamic acid. Cross-coupling reactions are pretty green from an atom economy standpoint, but they don't always use green reaction conditions. On the other hand, we will be running in water and using other green components, such as sodium carbonate and methanol.

Sodium carbonate reduces $PdCl_2$ to Pd(0) for the reaction, as well as functioning as a base.

Minimum safety standards for this experiment

1. Hot glass (and other equipment) looks the same as it does when it is cold. Since you are running the reaction at or above the boiling point of water, you must take care not to burn yourself.

2. Acrylic acid is a smelly irritant. Treat it with respect. It should be used only in the hood and measured by volume.

3. *p*-Iodoacetophenone is an irritant. Treat it with respect.

4. Sodium carbonate is mildly caustic but non-toxic. Wash your hands after handling.

5. Palladium dichloride is mildly corrosive, and because it is a fine powder it can cause respiratory difficulties if inhaled.

6. 1M Hydrochloric acid is corrosive but non-toxic. Wash your hands after handling.

7. Methanol is mildly toxic. Avoid skin contact where possible, and wash your hands after handling.

8. 4-Acetylcinnamic acid is an irritant. Treat it with respect.

Disposal

acrylic acid	Spillage should be cleaned with soap and water.
4-iodoacetophenone	Spillage should be swept up and placed in the trash.
sodium carbonate	Spillage should be swept up and placed in the trash. Solutions may be flushed down the sink.
palladium dichloride and palladium	Residues on filter paper should be placed in the wastebasket. Residues on glass may be flushed down the sink.
1M hydrochloric acid	Flush down the sink. Clean spills with paper towels, then place in the trash.
methanol	Flush down the sink. Clean spills with paper towels, then place in the trash.
aqueous waste from the reaction	Flush down the sink.
mother liquor from crystallization	Flush down the sink. Clean residues with acetone or isopropanol.
4-acetylcinnamic acid	Place in the wastebasket. Clean residues with acetone or isopropanol.

Procedure

You will be given a vial containing about 1.8 mg of palladium dichloride powder. After use, dispose of the vial in the trash.

This reaction runs at high temperatures, so take care when handling hot equipment!

1. In a 25-ml round-bottomed flask, combine the following: 4-iodoacetophenone (246 mg); sodium carbonate (318 mg); acrylic acid (0.100 mL, measured with an autopipet); and deionized water (5 ml). Add a magnetic stirring bar.

2. Add about 1 ml of DI water to the vial containing palladium dichloride powder to create a suspension. Add the suspension to the reaction mixture and rinse the vial into the reaction mixture with another 1 mL of water.

3. Attach a reflux condenser to the flask and heat **vigorously** at reflux for 60-75 minutes, maintaining rapid stirring. Afterward, remove the flask from the heat and allow it to cool to room temperature. Then filter off the catalyst by gravity filtration.

4. Add 1M HCl until the liquid mixture is acidic to litmus paper, and collect the precipitate by vacuum filtration.

5.   Place the filtered solids in a 25-ml Erlenmeyer flask and recrystallize using 50-50 methanol-water.

6.   Allow the product to dry for several days. Analyze your product for yield, and use melting point (lit.[1] 224-225°), NMR and IR to characterize it. Have you made *cis* or *trans* product? HINT: the coupling constants between vinylic hydrogen atoms that are cis to each other are about 5-10 Hz, while coupling constants between *trans* vinylic hydrogens are about 11-18 Hz.

   NOTE: 4-acetylcinnamic acid is not soluble in chloroform; you should dissolve it in acetone-d_6 for NMR. Remember that acetone-d_6 has a solvent peak around 2 ppm, which will be prominent if the solvent has not been freshly opened. Our acetone-d_6 contains 1% TMS, so you do not need to add TMS to your NMR sample.

For the report

Draw a detailed mechanism for the Heck coupling reaction that we are using. See Karty, Chapter 19, for a discussion.

Palladium-catalyzed cross-coupling reactions are typically considered "green." What is the atom economy for the coupling we performed? Does it seem particularly green? Why else would this reaction be considered "green"?

Pre-laboratory assignment

1.   Calculate the atom economy for the reaction we are performing. Don't forget that sodium carbonate is stoichiometric; the by-product is hydrogen iodide, which is neutralized by the base.

2.   During lab, we will discuss catalytic cycles and how they are commonly drawn.

3.   Are there any extraordinary hazards present in this experiment? Defend your answer. Consider not only the chemicals used, but also their amounts and the conditions.

---

[1] Cleland, G. H. *J. Org. Chem.* **1969**, *34*, 744–747

Nitration of methyl benzoate[1]

All aromatic compounds can be nitrated, using a mixture of nitric and sulfuric acids that generates the electrophile NO_2^+. If a substituent is already present on the ring, it will direct the new nitro group to a particular position. The mechanism of the formation of NO_2^+ and of its substitution onto the aromatic ring, and the directing effects of various substituents are discussed in your textbook.

In this experiment, we will nitrate methyl benzoate. The carbonyl group is the directing group. What product do we expect? You will have to prove that you have made the expected product.

NOTE: as a variation, you may be given one or two substituted methyl benzoates such as methyl 4-methoxybenzoate or methyl 4-nitrobenzoate. You will identify whether and where nitration takes place. Explain whether the result is consistent with the rule that the more active of two substituents directs the location of a third substitution.

Techniques used from Zubrick, 9th Ed: ice bath (page 169), vacuum filtration and recrystallization (Chapter 13)

Minimum Safety Standards for this experiment

1. The nitrating reagent we will use (a mixture of sulfuric and nitric acids) is concentrated, strong acid and can cause severe burns. Wear nitrile (blue) gloves – latex gloves will be burned by the acid – and avoid getting the stuff on your clothes.

2. Nitrated aromatic compounds are irritants and can cause skin rashes in those who are susceptible. Try to avoid skin contact and wash immediately after handling.

3. Methyl benzoate and its substituted analogs are irritants; their odor may make you sneeze. If it does, try not to drop anything!

4. Methanol is mildly toxic; ingestion can cause blindness. The amounts we use should be safe.

---

[1] Based on A.M. Schoffstall, B.A. Gaddis and M.L. Druelinger, *Microscale and Miniscale Organic Laboratory Experiments*, Boston: McGraw-Hill (2000), 296-297. Procedure courtesy of the Royal Society of Chemistry.

Disposal

Methyl benzoate and substituted methyl benzoates	Wipe up spills and place the paper towels used in a hood to dry. They may then be disposed of in the trash.
Methyl nitrobenzoate and substituted methyl nitrobenzoates	The quantities we produce may be disposed of in the trash.
Acid mixtures left over from the reaction	Flush down the sink.
Methanol	Flush down the sink.

Procedure

1. Weigh 2 g of methyl benzoate into a dry 50 ml Erlenmeyer flask.

2. Carefully transfer 1.5 mL of concentrated nitric acid into a dry test tube. Cool the nitric acid by partially immersing it in an ice-water bath before slowly adding, with swirling, 1.5 mL of concentrated sulfuric acid. Ensure thorough mixing. Allow this mixture to cool. This is the nitrating mixture.

3. Slowly add 4 mL of concentrated sulfuric acid to the methyl benzoate with swirling to ensure thorough mixing. Cool this mixture by partially immersing the flask in an ice-water bath.

4. Add the nitrating mixture very slowly (over about 15 min) to the contents of the Erlenmeyer flask. Stir or swirl the reaction mixture as each addition is made. During the addition keep the mixture in an ice bath, so that it remains below 6°C.

   NOTE: The nitrating mixture is very corrosive. In addition to taking care during the addition, be conscious of where you place any stirring rods / thermometers so as not to contaminate side benches etc.

5. Once addition is complete, allow the flask containing the reaction mixture to stand at room temperature for 15 min.

6. Carefully pour the reaction mixture onto a small amount (approx. 20 g) of crushed ice in a beaker. Stir the crushed ice as you add the reaction mixture. Solid methyl 3-nitrobenzoate will form.

7. Allow the ice to melt, then filter the mixture under suction. Wash the crude product with a little ice-cold water.

   Methyl 3-nitrobenzoate is insoluble in water but soluble in hot ethanol. Therefore, the crude material can be purified by recrystallization from a water/ethanol mixture.

8. Add 10 ml of DI water to the crude material in a small conical flask and warm the mixture to just below boiling. The methyl 3-nitrobenzoate will melt at this temperature but you will be able to see an "oily" substance within the water.

9. Slowly add hot reagent alcohol 1 ml at a time until the "oily" substance just dissolves.

10. Allow the mixture to cool to room temperature before cooling it further in an ice-water bath.

11. Filter the crystals under suction and allow them to dry.

12. Analyze your dried product by melting point, and, if necessary, IR and NMR. What are the expected melting points of the possible products? How many nitro groups are now attached to your methyl benzoate? What is the regiochemistry[1] of your product?

13. Discuss the mechanism of this reaction in your report, and explain the regiochemistry you found.

For the report
Draw a complete mechanism of the reaction, including the formation of the electrophile from nitric acid. Explain why you obtained the product that you did.

Pre-laboratory assignment
1. Draw the three possible mono-nitration products, and explain how you would distinguish them using NMR. What are their respective melting points?

2. Which of the three mono-nitration products do you expect to form?

3. Would you expect to add more than one nitro group? Why or why not?

4. Are there any extraordinary hazards present in this experiment? Defend your answer. Consider not only the chemicals used, but also their amounts and the conditions.

Nitration of ortho-vanillin[2]

All aromatic compounds can be nitrated, using a mixture of nitric and sulfuric acids that generates the electrophile NO_2^+. If a substituent is already present on the ring, it will direct the new nitro group to a particular position. The mechanism of the formation of NO_2^+ and of its substitution onto the aromatic ring, and the directing effects of various substituents are discussed in your textbook.

In this experiment, we will nitrate *ortho*-vanillin, 2-hydroxy-3-methoxybenzaldehyde. Which of three possible products do we expect? Why? You will have to prove that you have made the expected product.

The nitro group in the product is not shown on any specific carbon atom.

NOTE: The product of this reaction will be used to synthesize a photochromic compound next week.

Techniques used from Zubrick, 9th Ed: vacuum filtration and recrystallization (Chapter 13)

---

[1] That is, have you made methyl *ortho-*, *meta-* or *para*-nitrobenzoate? How does the new nitro group relate to the position of any other substituent that may have been present?

[2] Based on J.S. Griffiths, unpublished research at Stockton University, Galloway, NJ.

Minimum Safety Standards for this experiment

1. The nitrating reagent we will use (a mixture of acetic and nitric acids) is concentrated, strong acid and can cause severe burns. Wear nitrile (blue) gloves – latex gloves will be burned by the acid – and avoid getting the stuff on your clothes.

2. Glacial acetic acid is a corrosive, smelly irritant and must be kept in the hood. Wash after handling.

3. Nitrated aromatic compounds are irritants and can cause skin rashes in those who are susceptible. Try to avoid skin contact and wash immediately after handling.

4. Vanillin is an irritant, as is your product; its odor may make you sneeze.

Disposal

ortho-vanillin	Spills may be swept up and disposed of in the trash.
nitrated ortho-vanillin	Spills may be swept up and disposed of in the trash.
Acid mixtures left over from the reaction	Flush down the sink.
Methanol	Flush down the sink.

Procedure

1. Weigh 0.3-0.5 g of *ortho*-vanillin into a dry 25 ml round-bottomed flask. Add a magnetic stirring bar. Add 4-5 mL of glacial acetic acid and stir until the solid is dissolved.

2. After the solid is dissolved, add 1 mL of nitrating solution.[1] Note any color change. Stopper the flask and stir at room temperature for 10 minutes.

3. Add 5-6 mL of ice-cold water to your reaction mixture. Stopper and shake vigorously. Collect the solids by vacuum filtration.

4. Dry your product for at least 48 hours.

5. Analyze your dried product by melting point, and, if necessary, IR and NMR. What is the regiochemistry[2] of your product?

    -   2-hydroxy-3-methoxy-4-nitrobenzaldehyde,[3] m.p. 92-93°

    -   2-hydroxy-3-methoxy-5-nitrobenzaldehyde,[140] m.p. 140-142°

    -   2-hydroxy-3-methoxy-6-nitrobenzaldehyde,[4] m.p. (awaiting ILL)

---

[1] The nitrating solution is 15% nitric acid v/v in acetic acid, for example 15 mL nitric acid + 85 mL acetic acid.

[2] That is, is your nitro group on carbon 4, 5 or 6 of the *ortho*-vanillin ring? How does the new nitro group relate to the position of the three original substituents?

[3] V.K. Ahluwalia et al. *Intermediates for Organic Synthesis*, I.K. International Pvt. Ltd. (2010), pp. 65-66

[4] S. Kato, T. Morie, *J. Heterocyclic Chem.*, **1996**, *33*, 1171

6. Discuss the mechanism of this reaction in your report, and explain the regiochemistry you found. The WebMO companion exercise should help.

For the report

Draw a complete mechanism of the reaction, including the formation of the electrophile from nitric acid. Explain why you obtained the product that you did.

Pre-laboratory assignment

1. Draw the three possible nitration products, and explain how you would distinguish them using NMR.

2. Which of the three nitration products do you expect to form? Why?

3. Would you expect to add more than one nitro group? Why or why not?

4. Are there any extraordinary hazards present in this experiment? Defend your answer. Consider not only the chemicals used, but also their amounts and the conditions.

Synthesis of a photochromic compound

This procedure is based on unpublished research by J.S. Griffiths et al, The Richard Stockton College of New Jersey, Galloway NJ 08205.

In order for a compound to be colored, it must absorb light in the visible range; see Chapter 14 of Karty. Such compounds are always highly conjugated and often contain centers of positive or negative charge; one example is crystal violet, which you may have synthesized last semester.

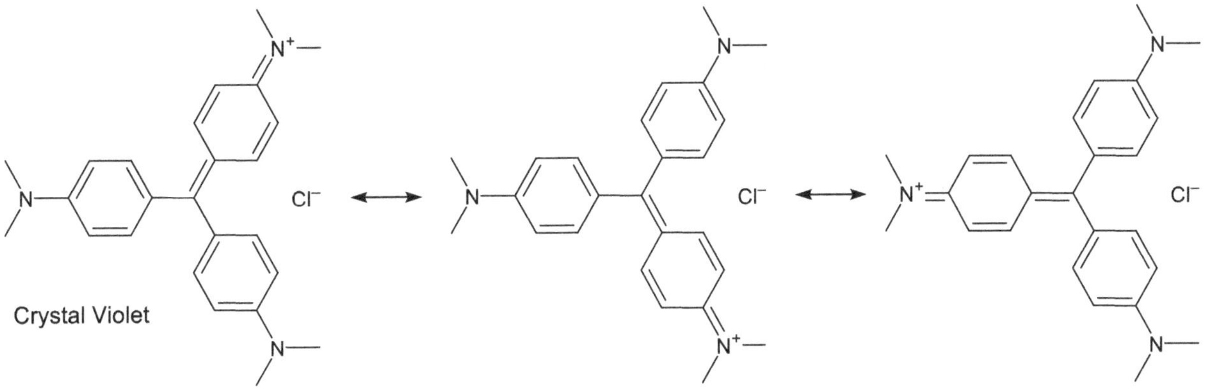

Some compounds can vary their electron distribution, and therefore their color, depending on the polarity of their solvent; such compounds are said to exhibit *solvatochromism*. One example is the merocyanine 1-methyl-4-[(oxocyclohexadienylidene)ethylidene]-1,4-dihydropyridine (MOED), which has two resonance structures, one more polar than the other:

Other compounds isomerize to highly colored forms when struck by light, heat, or influenced by a solvent. An example is the compound we will synthesize, 5'-chloro-1',3-dihydro-8-methoxy-1',3',3'-trimethyl-6-nitrospiro(2H-1-benzopyran-2,2'-2H-indole), which we will refer to as "chloro-BIPS."

Chloro-BIPS, a pale yellow or yellow-green compound, undergoes photochemical ring-opening to form a highly-colored merocyanine isomer. This isomer eventually re-cyclizes to the more stable BIPS structure.

"colorless" spiro form   colored merocyanine form

Draw curved arrows to show how the ring-opening takes place; note that photochemical excitation is required to change the double bond from cis to trans. Why is it able to thermally isomerize trans-to-cis for the re-cyclization reaction? Consider resonance forms!

Chloro-BIPS can be synthesized from 5-chloro-2-methylene-1,3,3-trimethylindoline – 2-methylene-1,1,3-trimethylindoline is called "Fisher's base" and so we will call this "chloro-Fisher's base" – and 2-hydroxy-3-methoxy-5-nitrobenzaldehyde ("5-nitro-*ortho*-vanillin" or "5-NOV").

chloro-Fisher's base   5-nitro-*ortho*-vanillin   chloro-BIPS
FW = 207.699   FW = 197.145   FW = 386.829

What is the mechanism of this reaction? (HINT: the initial step is addition of the methylene (CH$_2$) carbon to the aldehyde.)

Minimum Safety Standards for this experiment

1. Hot glass looks the same as cold glass. Be sure that a piece of glassware is cool enough to handle before picking it up.

2. We will use "reagent alcohol," a mixture that is 80% ethanol, 10% methanol and 10% 2-propanol. This liquid is moderately toxic and inflammable.

3. Chloro-Fisher's base, 5-nitro-ortho-vanillin and chloro-BIPS are all irritants. Treat them with respect and wash your hands after handling them.

Disposal

chloro-Fisher's base   This is provided as a solution that is about 0.1 grams per milliliter, in ethanol. Residues less than 1 mL may be flushed down the sink; larger amounts should be placed in halogenated organic waste.

5-nitro-ortho-vanillin   Small spills should be swept up and placed in the trash.

reagent alcohol   Flush down the sink.

chloro-BIPS   When you are finished, place your leftovers into the labeled bottle for use in CEM 105. Small spills should be swept up and placed in halogenated organic waste.

Leftover reaction mixture   The liquid remaining after filtration should be placed in halogenated organic waste.

Procedure

Note that you may be asked to synthesize your own 5-nitro-*ortho*-vanillin (5-NOV). If so, it will be done in a separate experiment.

1. Place 197 mg (1 mmol) of 5-NOV into a 25-mL round-bottomed flask with a magnetic stirring bar and 4 mL of reagent alcohol. Attach a condenser – water through the condenser will not be needed – and stir over the hot plate. Heat the hot plat until the 5-NOV dissolves completely.

2. The indoline ("chloro-Fisher's base") will be supplied as an alcohol solution that is approximately 0.5 moles per liter; the precise concentration will be given to you. Measure an amount that will deliver 1 mmol of indoline.[1] Add a little to your NOV solution and observe the immediate color change. Add the rest of the indoline solution, and stir for 20 minutes, heating enough for gentle reflux.

3. Allow the mixture to cool to room temperature; you may notice a yellowish precipitate.

4. Cool the mixture in an ice/water bath to finish crystallization. If you are not getting much product, you may need to add salt to your ice/water bath to bring the temperature below 0°. You can also make an ice/acetone bath to get the same effect.

5. Filter the solid product by vacuum filtration on a Hirsch funnel, and rinse your flask with a **small** amount of ice-cold alcohol. If a significant amount of solid appears in your filter flask, collect this solid and add it to your product.

6. At this point, you should dissolve **small** amounts of your product in several different solvents to observe solvatochromism. Irradiate your solutions for a few minutes under the soft UV lamp ("black light") provided and observe what happens.

7. Allow the product to dry for at least an hour before determining its final mass and taking a melting point (206-207°). Watch for something unusual happening upon melting: the drab spiro form will change into the highly colored merocyanine form.

---

[1] You should also adjust the amount based on how much 5-NOV you have; for example, if you have 1.1 mmol of 5-NOV, you need 1.1 mmol of indoline.

For the report

Draw a plausible mechanism for the reaction of chloro-Fisher's base with 5-nitro-*ortho*-vanillin.

Draw a plausible mechanism for the interconversion of the spiro form of chloro-BIPS and the merocyanine form.

Explain how the colors you observed indicate solvent polarity.

Pre-laboratory assignment

1. Draw a reasonable mechanism for BIPS ring-closing, from the merocyanine form to the spiro form.

2. Draw a reasonable mechanism of the reaction of the chloro-Fisher's base with 5-nitro-*ortho*-vanillin to form chloro-BIPS.

3. Are there any extraordinary hazards present in this experiment? Defend your answer. Consider not only the chemicals used, but also their amounts and the conditions.